T0291481

MATHEMATICAL MODELLING
IN ONE DIMENSION

AIMS Library Series

MATHEMATICAL MODELLING IN ONE DIMENSION

An Introduction via Difference and Differential Equations

JACEK BANASIAK

University of KwaZulu-Natal, South Africa
Technical University of Łódź

CAMBRIDGE
UNIVERSITY PRESS

CAMBRIDGE
UNIVERSITY PRESS

University Printing House, Cambridge CB2 8BS, United Kingdom

Cambridge University Press is part of the University of Cambridge.

It furthers the University's mission by disseminating knowledge in the pursuit of
education, learning and research at the highest international levels of excellence.

www.cambridge.org
Information on this title: www.cambridge.org/9781107654686

First published 2013

A catalogue record for this publication is available from the British Library

ISBN 978-1-107-65468-6 Paperback

Cambridge University Press has no responsibility for the persistence or accuracy
of URLs for external or third-party internet websites referred to in this publication,
and does not guarantee that any content on such websites is, or will remain,
accurate or appropriate.

Contents

Preface

Engineers, natural scientists and, increasingly, researchers and practitioners working in economics and other social sciences, use mathematical modelling to solve problems arising in their disciplines. There are at least two identifiable kinds of mathematical modelling. One involves translating the rules of nature or society into mathematical formulae, applying mathematical methods to analyse them and then trying to understand the implications of the obtained results for the original disciplines. The other kind is to use mathematical reasoning to solve practical industrial or engineering problems without necessarily building a mathematical theory for them.

This book is predominantly concerned with the first kind of modelling: that is, with the analysis and interpretation of models of phenomena and processes occurring in the real world. It is important to understand, however, that models only give simplified descriptions of real-life problems but, nevertheless, they can be expressed in terms of mathematical equations and thus can be solved in one way or another.

Mathematical modelling is a difficult subject to teach but it is what applied mathematics is all about. The difficulty is that there are no set rules and the understanding of the 'right' way to model can be only reached by familiarity with a number of examples. Therefore in this book we shall discuss a wide range of mathematical models referring to real life phenomena and introduce basic techniques for solving and interpreting the solutions of the resulting equations. It is, however, fair to emphasize that this is not a conventional textbook on mathematical modelling. We do not spend too much time on building the models but a central role is played by difference and differential equations and their analysis. However, within the space limitations, we try to describe the origin of the models and interpret the results of analysing them.

Nevertheless, let us briefly describe the full process of mathematical modelling. First, there must be a phenomenon of interest that one wants to describe or, more importantly, explain and make predictions about. Observation of this phenomenon allows one to make hypotheses about those quantities that are most relevant to the problem and what the relations between them are so that one can devise a hypothetical mechanism that describes the phenomenon. At this stage one has to decide how to quantify, or assign numbers to, the observations, e.g. whether the problem is to be set in absolute space-time or in relativistic setting, or whether time should be continuous or discrete, etc. The choice is not always obvious or unique but one needs to decide on a particular approach before one begins to build a model. The purpose of building the model is to formulate a description of the mechanism driving the phenomenon of interest in quantitative terms: that is, as mathematical equations which can be mathematically analysed. After that, it is necessary to interpret the solution, or any other information extracted from the equations, as statements about the original problem so that they can be tested against observations. Ideally, the model also leads to predictions which, if verified, serve as a further validation of the model. It is important to realize that modelling is usually an iterative procedure as it is very difficult to achieve a proper balance between the simplicity and meaningfulness of the model. Often the model turns out to be too complicated to yield itself to analysis or it is over-simplified so that there is insufficient agreement between actual experiment and the results predicted from the model. In both these cases one has to return to the first step of the modelling process to try to remedy the problem.

This first step in modelling is the most creative but also the most difficult, often involving a concerted effort of specialists in many diverse fields. Hence, as we said earlier, though we describe a number of models in detail, starting from first principles, the main emphasis of the course is on the later stages of the modelling process; that is, on analysing and solving the equations, interpreting their solutions in the language of the original problem and reflecting on whether the answers seem reasonable.

In most cases discussed here the model is a representation of a process; that is, it describes a change in the states of some system in time. This description could be discrete and continuous. The former corresponds to the situation in which we observe a system at regular finite time intervals, say, every second or every year, and relate the observed state of the system to the states at the previous instants. Such a system can be modelled by difference equations. In the continuous cases we treat time as a continuum allowing for observations of the system at any time. In such a case the model can express relations between the rates of change of various quantities rather than between the states at various times and, since the rates of change are given by derivatives, the model is represented by differential equations.

Furthermore, models can describe the evolution of a given system either in a non-interacting environment, such as a population of bacteria in a Petri dish, or else engaged in interactions with other systems. In the first case the model consists of a single equation and we say that the model is one-dimensional, while in the second case we have to deal with several (sometimes infinitely many) equations describing the interactions; then the model is said to be multi-dimensional.

In this book we only discuss one-dimensional models. Multi-dimensional models are planned to be the subject of a companion volume.

This book has been inspired and heavily draws on several excellent textbooks such as (Braun, 1983; Elaydi, 2005; Friedman and Littman, 1994; Glendinning, 1994; Strogatz, 1994), to mention but a few. However, I believe that the presented blend of discrete and continuous models and the combination of a detailed description of the modelling process with mathematical analysis of the resulting equations makes it different from any of them. I hope that the readers will find it fills a gap in the existing literature.

This book is based on lectures given, at various levels and for various courses, at the University of KwaZulu-Natal in Durban, at the African Institute of Mathematical Sciences in Muizenberg, South Africa, and at the Technical University of Łódź, Poland. My thanks go to several generations of students on whom I tested continually changing ideas of the course. Working with students allowed me to clarify many presentations and correct numerous mistakes. Finally, I am very grateful to our School Secretary, Dale Haslop, and my PhD student, Eddy Kimba Phongi, who, by combing through the final version of the book, helped to substantially reduce the number of remaining errors and thus to make the book more readable.

Jacek Banasiak
Durban, South Africa
October 2012

1
Mathematical toolbox

This book is about mathematical models coming from several fields of science and economics that are described by difference or differential equations. Therefore we begin by presenting basic concepts and tools from the theory of difference and differential equations, which will allow us to understand and analyse these models. The multitude of problems that can be dealt with using so few techniques is a testimony to the unifying power of mathematics.

1.1 Difference equations

Since difference equations are conceptually simpler, we begin with them. The reader should be aware that we present here a bare minimum of results that are necessary to analyse of the examples in this book. A comprehensive theory of difference equations can be found for example in (Elaydi, 2005).

We consider difference equations which can be written in the form

$$x_{n+k} = F(n, x_n, \ldots, x_{n+k-1}), \qquad n \in \mathbb{N}_0, \qquad (1.1)$$

where $k \in \mathbb{N}_0 = \{0, 1, 2, \ldots\}$ is a fixed number and F is a given function of $k + 1$ variables. Such an equation is called a difference equation of order k. If F does not explicitly depend on n, then we say that the equation is autonomous. Furthermore, if F depends linearly on $x_n, \ldots, x_{n+k-1}$, then we say that (1.1) is a linear equation. Otherwise, we say that it is nonlinear.

If we are given k initial values $x_1, \ldots, x_k$, then the term x_{k+1} is uniquely determined by (1.1) and then all other terms can be found by successive iterations. These terms form a sequence $(x_n)_{n \in \mathbb{N}_0}$ which we call *a solution* to (1.1). Thus the problems of existence and uniqueness of solutions, which play an essential role in the theory of differential equations, are here largely irrelevant. The problem, however, is to find a closed form of the solution; that is, a formula defining the terms of the sequence $(x_n)_{n \in \mathbb{N}_0}$ explicitly in terms of the variable n. While, in general, finding such an explicit solution is impossible, we shall discuss several cases when it can be accomplished. In more difficult situations we have to confine ourselves to qualitative analysis which will be discussed in Chapter 4.

1.1.1 First-order linear difference equations

The general first-order difference equation has the form

$$x_{n+1} = a_n x_n + g_n, \quad n \geq 0, \tag{1.2}$$

where $(a_n)_{n \in \mathbb{N}_0}$ and $(g_n)_{n \in \mathbb{N}_0}$ are given sequences. It is clear that using (1.2) we may calculate any element x_n provided we know only one initial point, so that we supplement (1.2) with an initial value x_0. It is easy to check, by induction, that the solution is given by

$$x_n = x_0 \prod_{k=0}^{n-1} a_k + \sum_{k=0}^{n-1} g_k \prod_{i=k+1}^{n-1} a_i, \tag{1.3}$$

where we adopt the convention that $\displaystyle\prod_{n}^{n-1} = 1$. Similarly, to simplify notation, we put $\displaystyle\sum_{k=j+1}^{j} = 0$.

Exercise 1.1 Show that if in (1.2) we have $a_n = a$ for all $n \geq 0$, then (1.3) takes the form

$$x_n = a^n x_0 + \sum_{k=0}^{n-1} a^{n-k-1} g_k. \tag{1.4}$$

If, moreover $g_n = g$ for $n \geq 0$, then

$$x_n = \begin{cases} a^n x_0 + g\frac{a^n-1}{a-1} & \text{if} \quad a \neq 1, \\ x_0 + gn & \text{if} \quad a = 1. \end{cases} \tag{1.5}$$

1.1.2 Linear difference equations of higher order

Though the book mainly is concerned with equations of first order, in some examples we will need solutions to higher-order linear equations with constant coefficients; that is, equations of the form

$$x_{n+k} + a_1 x_{n+k-1} + \cdots + a_k x_n = 0, \qquad n \in \mathbb{N}_0, \tag{1.6}$$

where k is a fixed number, called the order of the equation, and $a_1, \ldots, a_k$ are known numbers. This equation determines the values of x_m, $m > k$, by k preceding values. Thus, we need k initial values $x_0, x_1, \ldots, x_{k-1}$ to start iterations. The general theory of such equations requires tools from linear algebra, which are beyond the scope of this book, see (Elaydi, 2005). Therefore we only will present basic results which easily can be checked to hold true in particular examples.

To find the general solution to (1.6), we build the so-called *characteristic equation*

$$\lambda^k + a_1 \lambda^{k-1} + \cdots + a_k = 0. \tag{1.7}$$

If this equation has k distinct roots $\lambda_1, \ldots, \lambda_k$, then the general solution is given by

$$x_n = C_1 \lambda_1^n + \cdots + C_k \lambda_k^n, \quad n \geq k, \tag{1.8}$$

where $C_1, \ldots, C_k$ are constants that are to be determined so that $(y_n)_{n \in \mathbb{N}_0}$ satisfies the initial conditions for $n = 0, \ldots k - 1$. If, however, there is a multiple root, say λ_i, of multiplicity n_i, then in the expansion (1.8) we must use n_i terms $\{\lambda_i^n, n\lambda_i^n, \ldots, n^{n_i-1}\lambda_i^n\}$.

1.1.3 Nonlinear equations

As we said earlier, most difference equations cannot be solved explicitly. In some cases, however, a smart substitution could reduce them to a simpler form. In this subsection we present two classes of solvable nonlinear equations, which will be used later. Some other cases are discussed in Section 2.4.

The homogeneous Ricatti equation. Consider the equation

$$x_{n+1}x_n + a_n x_{n+1} + b_n x_n = 0, \quad n \in \mathbb{N}_0, \tag{1.9}$$

where $(a_n)_{n\in\mathbb{N}_0}$ and $(b_n)_{n\in\mathbb{N}_0}$ are given sequences with non-zero elements. Then the substitution

$$y_n = \frac{1}{x_n}$$

transforms (1.9) into

$$b_n y_{n+1} + a_n y_n + 1 = 0, \tag{1.10}$$

which is a first-order linear equation. We note that in the above transformation we had to assume $x_n \neq 0$. If, however, $x_n = 0$ for some n, then $x_m = 0$ for $m > n$.

The inhomogeneous Ricatti equation. The inhomogeneous Riccati equation is

$$x_{n+1}x_n + a_n x_{n+1} + b_n x_n = c_n, \quad n \in \mathbb{N}_0, \tag{1.11}$$

where $(a_n)_{n\in\mathbb{N}_0}$, $(b_n)_{n\in\mathbb{N}_0}$ and $(c_n)_{n\in\mathbb{N}_0}$ are given sequences. Upon the substitution

$$x_n = \frac{y_{n+1}}{y_n} - a_n,$$

it becomes

$$\left(\frac{y_{n+2}}{y_{n+1}} - a_{n+1}\right)\left(\frac{y_{n+1}}{y_n} - a_n\right) + a_n\left(\frac{y_{n+2}}{y_{n+1}} - a_{n+1}\right)$$
$$+ b_n\left(\frac{y_{n+1}}{y_n} - a_n\right) = c_n.$$

Simplifying, we obtain the second-order linear equation

$$y_{n+2} + (b_n - a_{n+1})y_{n+1} - (c_n + a_n b_n)y_n = 0. \tag{1.12}$$

In particular, if the sequences $(a_n)_{n\in\mathbb{N}_0}$, $(b_n)_{n\in\mathbb{N}_0}$ and $(c_n)_{n\in\mathbb{N}_0}$ are constant, then the above equation is explicitly solvable by the method described in Section 1.1.2.

1.2 Differential equations – an introduction

The present book is mostly about applying differential equations to concrete models, thus we refer the reader to dedicated texts, such as (Braun, 1983; Glendinning, 1994; Schroers, 2011; Strogatz, 1994), to learn more about the theory of differential equations. However, to make the presentation self-consistent, we provide some basic facts and ideas.

In this book we shall be solely concerned with ordinary differential equations (ODEs) that can be written in the form

$$y^{(n)} = F(t, y, y', \ldots, y^{(n-1)}) = 0, \tag{1.13}$$

where F is a given scalar function of $n + 1$ variables and $y^{(k)}$, for $k = 1, \ldots, n$, denotes the derivative of order k with respect to t. For lower order derivatives we will use the more conventional notation $y^{(1)} = y', y^{(2)} = y''$, etc. As with the difference equations, we say that (1.13) is autonomous if F does not depend on t and it is linear if F is linear in $y, y', \ldots, y^{(n-1)}$. The order of the equation is the order of the highest derivative appearing in it.

To solve the ODE (1.13) means to find an n-times continuously differentiable function $y(t)$ such that for any t (from some interval), (1.13) becomes an identity. Thus, if we are given a function y, it is easy to check whether it is a solution of (1.13) or not. However, in contrast to difference equations, finding a solution to (1.13) is a difficult, and often impossible, task. A quick reflection brings to mind three questions relevant to solving a differential equation:

(i) can we be sure that a given equation possesses a solution at all?

(ii) if we know that there is a solution, are there systematic methods for finding it?

(iii) having found a solution, can we be sure that there are no other solutions?

Question (i) is usually referred to as the **existence problem** for differential equations, and Question (iii) as the **uniqueness problem**. Unless we deal with very simple situations, these two questions should be addressed before attempting to find a solution.

After all, what is the point of trying to solve an equation if we do not know whether the solution exists, or whether the one we found is unique. Let us discuss briefly Question (i) first. Roughly speaking, we can come across the following situations:

(a) no function exists which satisfies the equation;
(b) the equation has a solution but no one knows what it looks like;
(c) the equation can be solved in a closed form.

Case (a) is not very common in mathematics and it should never happen in mathematical modelling. Indeed, if a given equation was an exact reflection of a real life phenomenon, then the fact that this phenomenon exists would ensure that this equation can be solved. However, models are imperfect reflections of the reality and therefore it may happen that in the modelling process we missed some crucial facts, rendering the final equation unsolvable. Thus, establishing solvability of the equation constructed in the modelling process serves as an important first step in validating the model. Unfortunately, these problems are usually very difficult and require quite advanced mathematics that is beyond the scope of this course. We shall, however, provide basic theorems pertaining to this question that are sufficient for the discussed problems.

Case (b) may look somewhat enigmatic but, as we said above, there are advanced theorems allowing us to ascertain the existence of solutions without actually displaying them. Actually, many of the most interesting equations appearing in applications do not have known explicit solutions. It is important to realize that even if we do not know a formula for the solution, the fact that one does exist means we can find its numerical or graphical representation to any reasonable accuracy. Also, very often we can find important features of the solution without knowing its explicit formula. These features include e.g., long time behaviour; that is, whether it settles at a certain equilibrium value or oscillates, whether it is monotonic or periodic, etc. These questions will be studied in the final part of the book.

Some examples, when the situation described in (c) occurs and which thus also partially address Question (ii), are discussed in Section 1.3 below.

Having dealt with Questions (i) and (ii) let us move to the problem of uniqueness. Typically (1.13) determines a family of solutions, parametrised by several constants, rather than a single function. Such a class is called the *general solution* of the equation. By imposing an appropriate number of *side conditions* we specify the constants thus obtaining a *special solution* – ideally one member of the class.

A side condition may take all sorts of forms, such as 'at $t = 15$, y must have the value of 0.4' or 'the area under the curve $y = y(t)$ between $t = 0$ and $t = 24$ must be 100'. Very often, however, it specifies the initial value $y(0)$ of the solution and the derivatives $y^{(k)}(0)$ for $k = 1, \ldots, n - 1$. In this case the side conditions are called the *initial conditions*. Problems consisting of (1.13) with initial conditions are called *initial value problems* or *Cauchy problems*

1.3 Some equations admitting closed form solutions

In this section we shall provide a brief overview of methods for solving differential equations which will appear in this book. This shows that in some situations the answer to Question (ii) of the previous section is affirmative. It is important to understand, however, that there is a deeper theory behind each method and due caution should be exercised when applying the formulae listed below, see (Braun, 1983; Schroers, 2011; Strogatz, 1994).

1.3.1 Separable equations

Separable equations are equations which can be written as

$$y' = g(t)h(y), \tag{1.14}$$

where g and h are known functions. Constant functions $y \equiv \bar{y}$, such that $h(\bar{y}) = 0$, are solutions to (1.14). They are called *stationary or equilibrium solutions*.

To find the general solution, we assume that $h(y)$ is finite and nowhere zero, and divide both sides of (1.14) by $h(y)$ to get

$$\frac{1}{h(y)}y' = g(t). \tag{1.15}$$

Denoting $H(y) = \int dy/h(y)$, (1.15) can be written as

$$(H(y(t)))' = g(t).$$

Integrating, we obtain the solution in the implicit form,

$$H(y(t)) = \int g(t)dt + c, \tag{1.16}$$

where c is an arbitrary constant. Since, by assumption, $H'(y) = h^{-1}(y) \neq 0$, we can use the inverse function theorem (Courant and John, 1999) to claim that the function H is locally invertible and thus the explicit solution can be found, at least locally, as

$$y(t) = H^{-1}\left(\int g(t)dt + c\right), \tag{1.17}$$

with c depending on the side conditions.

1.3.2 First-order linear differential equations

The general *first-order linear differential equation* is of the form

$$y' + a(t)y = b(t), \tag{1.18}$$

where a and b are known continuous functions of t. One method of solving (1.18) is to multiply both sides of (1.18) by the so-called *integrating factor* μ which is a solution to

$$\mu' = \mu a(t),$$

i.e., $\mu(t) = e^{\int a(t)dt}$. Then

$$\mu(t)y' + \mu(t)a(t)y = \mu(t)b(t)$$

can be written as

$$(\mu(t)y(t))' = \mu(t)b(t),$$

and thus

$$y(t) = \frac{1}{\mu(t)} \left(\int \mu(t)b(t)dt + c \right) \tag{1.19}$$

$$= \exp\left(-\int a(t)dt \right) \left(\int b(t)\exp\left(\int a(t)dt \right) dt + c \right),$$

where c is a constant of integration which is to be determined from the initial conditions. It is worthwhile noting that the solution is the sum of the general solution to the homogeneous equation (that is, with $b(t) \equiv 0$),

$$c\exp\left(-\int a(t)dt \right),$$

and a particular solution to the full equation (1.18).

1.3.3 Equations of homogeneous type

A differential equation that can be written in the form

$$y' = f\left(\frac{y}{t}\right), \tag{1.20}$$

where f is a function of the single variable $z = y/t$ is said to be of *homogeneous type*. To solve (1.20), let us make the substitution

$$y = tz, \tag{1.21}$$

where z is the new unknown function. Then, by the product rule,

$$y' = z + tz'$$

and (1.20) becomes

$$tz' = f(z) - z. \tag{1.22}$$

Equation (1.22) is a separable equation and so it can be solved as in Section 1.3.1.

1.3.4 Equations that can be reduced to first-order equations

Some higher-order equations can be reduced to first-order equations. We shall discuss two such cases for second-order equations.

Equations that do not contain the unknown function. If we have an equation of the form

$$F(t, y', y'') = 0, \qquad (1.23)$$

then the substitution $z = y'$ reduces this equation to the first-order equation

$$F(t, z, z') = 0. \qquad (1.24)$$

If

$$z = \phi(t, C)$$

is the general solution to (1.24), where C is an arbitrary constant, then y is the solution of

$$y' = \phi(t, C),$$

so that

$$y(t) = \int \phi(t, C)dt + C_1.$$

Equations that do not contain the independent variable. Let us consider the equation

$$F(y, y', y'') = 0, \qquad (1.25)$$

that does not involve the independent variable t. Such an equation also can be reduced to a first-order equation as long as $y' \neq 0$; that is, if there are no turning points of the solution. Then the derivative y' locally is a function of y; that is, we can write $y' = g(y)$ for some function g. Indeed, by the inverse function theorem, see (Courant and John, 1999), the function $y = y(t)$ is locally invertible provided $y' \neq 0$ and, writing $t = t(y)$, we can define $g(y) = y'(t(y))$. Using the chain rule we obtain

$$y'' = \frac{d}{dt}y' = \frac{dg}{dy}\frac{dy}{dt} = y'\frac{dg}{dy} = g(y)\frac{dg}{dy}. \qquad (1.26)$$

Substituting (1.26) into (1.25) gives a first-order equation with y as an independent variable,

$$F\left(y, g, g\frac{dg}{dy}\right) = 0. \qquad (1.27)$$

If $g(y) = \phi(y, C)$ is the solution to (1.27), then y satisfies the separable equation

$$\frac{dy}{dt} = \phi(y, C)$$

which can be solved by (1.17).

The above procedure can be best explained by interpreting t as time, y as the distance travelled by a particle moving with velocity y' and acceleration y''. If the particle does not reverse the direction of motion ($y' = 0$ at any turning point!), then its speed can be expressed as a function of the distance instead of time. This is precisely what we have done above.

1.4 The Cauchy problem – existence and uniqueness

In most cases in this book we will be concerned with the first-order Cauchy problem

$$y' = f(t, y), \qquad y(t_0) = y_0, \tag{1.28}$$

where t_0 and y_0 are some given numbers. For unspecified or very complicated functions f, none of the methods described in Section 1.3 is applicable and we have to resort to an abstract approach. The following general existence result is known as the *Peano theorem*, see (Robinson, 2001).

Theorem 1.2 (Peano) *If the function f in (1.28) is continuous in some neighbourhood of the point (t_0, y_0), then the problem (1.28) has at least one solution in some interval (t_1, t_2) containing t_0.*

Since the equation in (1.28) is a first-order equation and we have one initial condition, one would expect that it would be uniquely solvable. Unfortunately, in general this is not so as demonstrated in the following example.

Example 1.3 The Cauchy problem

$$y' = \sqrt{y}, \quad t > 0,$$
$$y(0) = 0,$$

has at least two solutions: $y \equiv 0$ and $y(t) = \frac{1}{4}t^2$.

Fortunately, there is a large class of functions f for which (1.28) does have exactly one solution. This result is Picard's theorem which we state below.

Theorem 1.4 (Picard) *Assume that there is a rectangle $\mathcal{R}$: $|t - t_0| \leq a, |y - y_0| \leq b$ for some $a, b > 0$, such that the function f in (1.28) is continuous in $\mathcal{R}$ and satisfies the Lipschitz condition with respect to y there; i.e. there exists $0 \leq L < +\infty$ such that for all $(t, y_1), (t, y_2) \in \mathcal{R}$*

$$|f(t, y_1) - f(t, y_2)| \leq L|y_1 - y_2|. \tag{1.29}$$

Let

$$M := \max_{(t,y) \in R} |f(t, y)|$$

and define $\alpha = \min\{a, b/M\}$. Then the initial value problem (1.28) has exactly one solution defined at least on the interval $t_0 - \alpha \leq t \leq t_0 + \alpha$.

Remark 1.5 If f is such that its partial derivative with respect to y, namely f_y, is bounded in $\mathcal{R}$, then (1.29) is satisfied.

Picard's theorem gives local existence and uniqueness; that is, for any point (t_0, y_0) around which the assumptions are satisfied, there is an interval over which there is only one solution of the given Cauchy problem. This interval can be very small, smaller than $[t_0 - a, t_0 + a]$ on which f satisfies the assumptions of the theorem. However, using Picard's theorem we can glue solutions together to obtain a solution defined on a possibly larger interval. More precisely, if $y(t)$ is a solution to (1.28) defined on an interval $[t_0 - \alpha, t_0 + \alpha]$ and $(t_0 + \alpha, y(t_0 + \alpha))$ is a point around which the assumptions of Picard's theorem are satisfied, then there is a solution passing through this point, defined on some interval $[t_0 + \alpha - \alpha', t_0 + \alpha + \alpha']$, $\alpha' > 0$. By uniqueness, these two solutions constitute a solution to the original Cauchy problem, defined at least on $[t_0 - \alpha, t_0 + \alpha + \alpha']$. By continuing this process we obtain a solution defined on the *maximal interval of existence* $[t_0, t_0 + \alpha^*)$. In other words, $[t_0, t_0 + \alpha^*)$ is the (forward) maximal interval of existence for a solution $y(t)$ to (1.28) if there is no solution $y_1(t)$ on an interval $[t_0, t_0 + \alpha^+)$, where $\alpha^+ > \alpha^*$, satisfying $y(t) = y_1(t)$

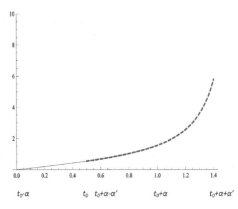

Figure 1.1 Extension of the solution from $[t_0 - \alpha, t_0 + \alpha]$ to $[t_0 - \alpha, t_0 + \alpha + \alpha']$

for $t \in [t_0, t_0 + \alpha^*)$. We can also consider backward intervals trying to extend the solution for $t < t_0$. We note that the forward (backward) maximal interval of existence is open from the right (left). Assume now that f satisfies the assumptions of the Picard theorem on each rectangle $\mathcal{R} \subset \mathbb{R}^2$ (possibly with different Lipschitz constants). The important question of whether the solution exists for all times; that is, whether $\alpha^* = \infty$, is addressed in the following theorem.

Theorem 1.6 *If we assume that f in* (1.28) *satisfies the assumptions of the Picard theorem on any rectangle $\mathcal{R} \subset \mathbb{R}^2$, then $[t_0, t_0 + \alpha^*)$ is a finite forward maximal interval of existence of $y(t)$ if and only if*

$$\lim_{t \to t_0 + \alpha^*} |y(t)| = \infty. \tag{1.30}$$

In other words, a solution to an equation with a regular right hand side either exists for all times, or blows up (becomes infinite) in a finite time. A classical proof of this result can be found in e.g., (Robinson, 2001). In Section 4.1.1 we derive it from the general description of the dynamics of a scalar equation.

These theorems are of great theoretical and practical importance. We discuss some applications below.

Example 1.7 We consider the first-order linear differential equation

$$y' = ay. \tag{1.31}$$

This is an example of a separable equation, discussed in Section 1.3.1. Thus, if $y(t) \neq 0$ for any t, then (1.16) gives

$$\ln |y(t)| = at + c_1,$$

where c_1 is an arbitrary constant of integration. Taking exponentials of both sides yields

$$|y(t)| = \exp(at + c_1) = c_2 \exp(at),$$

where c_2 is an arbitrary positive constant: $c_2 = \exp c_1 > 0$. We would like to discard the absolute value bars at $y(t)$. To do this, observe that in the derivation we required that $y(t) \neq 0$ for any t, thus y, being a continuous function, must be of a constant sign. Hence,

$$y(t) = \pm c_2 \exp(at) = c_3 \exp(at) \tag{1.32}$$

where c_3 can be either positive or negative.

Are these all possible solutions to (1.31)? Solution (1.32) was derived under provision that $y \neq 0$. We clearly see that $y \equiv 0$ is a solution to (1.31) but, fortunately, this solution can be incorporated into (1.32) by allowing c_3 to be zero.

However, we still have not ruled out the possibility that the solution can cross the x-axis at one or more points. We cannot use here (1.32), as it was derived under the assumption that this cannot happen. Thus we must resort to the Picard theorem. First of all we note that the function $f(t, y)$ is given by $f(t, y) = ay$ and $|f(t, y_1) - f(t, y_2)| = a|y_1 - y_2|$ so that f satisfies assumptions of the Picard theorem on any closed rectangle $\mathcal{R} \subset \mathbb{R}^2$ with Lipschitz constant $L = a$. If there was a solution satisfying $y(t_0) = 0$ for some t_0 then, from the uniqueness part of the Picard theorem, this solution should be identically zero, as $y(t) \equiv 0$ is a solution to this problem. In other words, if a solution to (1.31) is zero at some point, then it is identically zero.

In many applications it is important to know that if the initial condition is non-negative, then so is the corresponding solution.

For instance, in population dynamics the solution of a differential equation is the number of individuals in a population and thus typically a model which admits negative solutions for positive data cannot be correct. Below we will see how a variant of the argument used in Example 1.7 can be used to establish that the above property holds for a class of differential equations.

Example 1.8 Let us consider the Cauchy problem

$$y' = yg(t, y), \qquad y(t_0) = y_0, \qquad (1.33)$$

where g is a continuous function on some neighbourhood of $[t_0 - a, t_0 + a) \times (y_0 - b, y_0 + b]$, $a, b > 0$ and let y be a solution to (1.33), defined on an interval $[t_0, t_1)$, with some $t_0 < t_1 < t_0 + b$. Then $y_0 > 0$ implies $y(t) > 0$. To prove this we employ a trick which is often used to prove properties of solutions of differential equations. It uses the fact that since we know that the solution exists, we can treat some of the unknowns y in (1.33) as known functions, thus reducing the complexity of the equation. To explain this in detail, we note that if y is a solution, then we have the identity

$$y'(t) \equiv y(t)g(t, y(t)), \quad t \in [t_0, t_1).$$

Since, in principle, y is a known function, then we can define $\tilde{g}(t) = g(t, y(t))$ and re-write (1.33) as the linear Cauchy problem

$$y' = y\tilde{g}(t), \qquad y(t_0) = y_0, \qquad (1.34)$$

where $\tilde{g}$ is a continuous function. Then, as in Example 1.7, we see that $y\tilde{g}(t)$ satisfies the assumptions of the Picard theorem (note that $\tilde{g}$ only is a function of t). Hence, again arguing as in Example 1.7, we see that if $y_0 > 0$, then

$$y(t) = y_0 \exp \int_{t_0}^{t} \tilde{g}(s)ds$$

and clearly $y(t) > 0$ for all t for which it is a solution to (1.33).

To illustrate applications of Theorem 1.6, let us consider some examples.

Exercise 1.9 Consider the Cauchy problem

$$y' = 1 + y^2, \quad y(0) = 0, \qquad (1.35)$$

Show that the Picard theorem ensures the existence of a unique solution to (1.35) for $|t| \leq \frac{1}{2}$.

Next, show that the explicit solution to (1.35) is $y(t) = \tan t$ which is defined on $|t| < \pi/2$.

This exercise shows that, in general, the Picard theorem does not give the best possible answer – that is why it is sometimes referred to as a local existence theorem. On the other hand, the right hand side of the equation satisfies the assumptions of the Picard theorem everywhere and thus the solution ends its existence at finite $t = +\pi/2$ with a 'bang', in accordance with Theorem 1.6.

Let us assume that (1.28) describes the evolution of a population. Furthermore, let its right hand side satisfy the assumptions of the Picard theorem and let positive initial data produce positive solutions, as in Example 1.8. Then, according to Theorem 1.6, the population either exists for all time or suffers an explosion in a finite time. In other words, within this framework the only way for a population to exist only for a finite time is to blow up. Since, clearly, real populations can become extinct (that is, reach zero) in finite time, the above results are not satisfactory. One possible remedy is to consider equations whose solutions can change sign (e.g., for which $y \equiv 0$ is not a solution). Then we can interpret the first time the solution becomes zero as the time of the extinction of the population. Another possibility is to weaken the assumption on the right hand side of (1.28) so that Theorem 1.6 will cease to be applicable. Such a scenario is described in the next example.

Example 1.10   Consider the following Cauchy problem

$$y' = -\frac{1}{2y}, \qquad y(0) = 1.$$

This is a separable equation which can be transformed to

$$-1 = 2yy' = (y^2)'$$

and, upon integration, yields the solution $y^2 = -t + C$. Hence, using the initial condition $y(0) = 1 > 0$, we obtain

$$y(t) = \sqrt{1 - t}, \quad t < 1.$$

Therefore, the solution $y(t)$ exists only on the open interval $t < 1$,

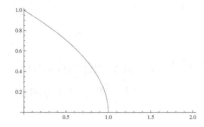

Figure 1.2 The graph of the solution in Example 1.10

however, contrary to Exercise 1.9, the solution simply vanishes at the endpoint. We note that this does not violate Theorem 1.6 as at the point $(1,0)$, where the solution vanishes, the right hand side is not Lipschitz continuous.

It is equally important to have easy-to-use criteria ensuring that solutions of (1.28) are defined for all $t \in \mathbb{R}$. One of the most often used criteria is that f is *globally Lipschitz continuous*: that is, that f satisfies assumptions of the Picard theorem on $\mathbb{R}^2$ with the constant L in (1.29) independent of the rectangle $\mathcal{R}$. In such a case the solution y exists for all $t \in \mathbb{R}$. Usually this result is proved by combining Theorem 1.6 with the so-called Gronwall lemma, see e.g., (Robinson, 2001). Later, in Remark 4.4, we shall see how it follows from the general description of the dynamics of a scalar equation.

2

Basic difference equations models and their analysis

In this chapter we first introduce discrete mathematical models of phenomena happening in the real world. We begin with some explanatory words. Apart from the simplest cases such as the compound interest equation, where the equation is a mathematical expression of rules created by ourselves, the mathematical model attempts to find equations describing events happening according to their own rules, our understanding of which is far from complete. At best, the model can be an approximation of the real world. This understanding guides the way in which we construct the model: we use the principle of economy (similar to the Ockham razor principle) to find the simplest equation which incorporates all relevant features of the modelled events. Such a model is then tested against experiment and only adjusted if we find that its description of salient properties of the real phenomenon we try to model is unsatisfactory.

This explains why we often begin modelling by fitting a linear function to the data and why such linear, or only slightly more complicated, models are commonly used, although everybody agrees that they do not properly describe the real world. The reason is that often they supply sufficient, if not exact, answers at a minimal cost. One must remember, however, that using such models is justified only if we understand their limitations and that, if necessary, are ready to move in with more fine-tuned ones.

2.1 Difference equations of financial mathematics

2.1.1 Compound interest

Compound interest is relevant to loans or deposits made over long periods. The interest is added to the initial sum at regular intervals, called the conversion periods, and the new amount, rather than the initial one, is used for calculating the interest for the next conversion period. The fraction of a year occupied by the conversion period is denoted by α. Thus the conversion period of 1 month is given by $\alpha = 1/12$. Instead of saying that the conversion period is 1 month we also say that the interest is compounded monthly.

For an annual interest rate of $p\%$ and the conversion period equal to α, the interest earned for the period is equal to $\alpha p\%$ of the amount deposited at the start of the period; that is,

$$\left\{\begin{array}{c} \text{amount} \\ \text{deposited} \\ \text{after } k+1 \\ \text{conversion} \\ \text{periods} \end{array}\right\} = \left\{\begin{array}{c} \text{amount} \\ \text{deposited} \\ \text{after } k \\ \text{conversion} \\ \text{periods} \end{array}\right\} + \frac{\alpha p}{100}\left\{\begin{array}{c} \text{amount} \\ \text{deposited} \\ \text{after } k \\ \text{conversion} \\ \text{periods} \end{array}\right\}.$$

To express this as a difference equation, for each k let S_k denote the amount on deposit after k conversion periods. Thus

$$S_{k+1} = S_k + \frac{\alpha p}{100}S_k = S_k\left(1 + \frac{\alpha p}{100}\right) \qquad (2.1)$$

which is a simple first-order linear difference equation. Using (1.5) with $g = 0$ (or simply noting that S_k follows the geometric progression), we get

$$S_k = \left(1 + \frac{\alpha p}{100}\right)^k S_0, \qquad (2.2)$$

which is called the compound interest formula. If we want to measure time in years, then $k = t/\alpha$, where t is the number of years. Then (2.2) takes the form

$$S_t = \left(1 + \frac{\alpha p}{100}\right)^{t/\alpha} S_0. \qquad (2.3)$$

It is worth introducing here the concept of effective interest rate.

First we note that in (2.3), with $S_0 = 1$,

$$S_1 = \left(1 + \frac{\alpha p}{100}\right)^{1/\alpha} = 1 + \frac{p}{100} + \cdots > 1 + \frac{p}{100},$$

so if the interest is compounded several times per year, the increase in savings is bigger than if it was compounded annually. This is the basis of defining the effective interest rate p_{eff} (relative to the conversion period), often used in banks' commercials. Namely

$$1 + p_{\text{eff}} = \left(1 + \frac{\alpha p}{100}\right)^{1/\alpha}, \tag{2.4}$$

that is, p_{eff} is the interest rate which, compounded annually, would give the same return as the interest p compounded with conversion period α.

Exercise 2.1   In some banks the interest depends on the amount in the account. Derive the difference equation that describes the growth of your deposit in such a case, taking the conversion period α and interest rate $p(N)$.

Exercise 2.2   Modify (2.1) to describe the situation, when you have to pay an administrative fee of $q(N)$ every time the interest is added.

2.1.2 Loan repayment

A slight modification of the above argument can be used to find the equation governing loan repayments. The scheme described below usually is used for the repayment of house or car loans. Repayments are made at regular intervals in equal amounts to reduce the loan and to pay the interest on the amount still owing.

It is supposed that the compound interest at $p\%$ is charged on the outstanding debt, with the conversion period equal to the same fraction α of the year as the period between the repayments. Between the payments, the debt increases because of the interest charged on the debt still outstanding after the last repayment.

Hence,

$$\left\{ \begin{array}{c} \text{debt after} \\ k+1 \text{ payments} \end{array} \right\} = \left\{ \begin{array}{c} \text{debt after} \\ k \text{ payments} \end{array} \right\}$$

$$+ \left\{ \begin{array}{c} \text{interest} \\ \text{on this debt} \end{array} \right\} - \{\text{payment}\},$$

where it is assumed that the payment is made at the end of each conversion period. To write this as a difference equation, let D_0 be the initial debt to be repaid and, for each k, let the outstanding debt after the kth repayment be D_k. Then

$$D_{k+1} = D_k + \frac{\alpha p}{100} D_k - R = D_k \left(1 + \frac{\alpha p}{100}\right) - R.$$

We note that if the instalment was paid at the beginning of the conversion period, the equation would take a slightly different form

$$D_{k+1} = D_k - R + \frac{\alpha p}{100}(D_k - R) = (D_k - R)\left(1 + \frac{\alpha p}{100}\right).$$

The reason for the change is that the interest is calculated from the debt D_k reduced by the payment R done at the beginning of the conversion period.

These equations fit into theory presented in Section 1.1.1. To simplify notation, we put $r = \alpha p/100$ and, by (1.5), we obtain the solution

$$D_k = (1+r)^k D_0 - R \sum_{i=0}^{k-1} (1+r)^{k-i-1}$$

$$= (1+r)^k D_0 - \left((1+r)^k - 1\right)\frac{R}{r}. \tag{2.5}$$

Exercise 2.3 Show that the formula for the monthly instalment on a loan D_0 to be repaid in n instalments is

$$R = \frac{r D_0}{1 - (1+r)^{-n}}. \tag{2.6}$$

Hint. Solve (2.5) subject to $D_n = 0$.

Calculate the instalments on a mortgage of $200000 to be repaid over 20 years in monthly instalments at the annual interest rate of 3%.

2.1.3 Gambler's ruin

The next money-related problem involves a different type of modelling with roots in probability theory. Problems of this type are common in the theory of Markov chains, see e.g. (Feller, 1968).

A gambler plays a sequence of games against an adversary. The probability that the gambler wins \$1 in any given game is q and the probability of him losing \$1 is $1 - q$. He quits the game if he either wins a prescribed amount of \$$N$, or loses all his money; in the latter case we say that he has been ruined. Let p_n denote the probability that the gambler will be ruined if he starts gambling with \$$n$. We build the difference equation satisfied by p_n, using the following argument. Firstly, note that we can start observations at any moment; that is, the probability of the gambler being ruined with \$$n$ at the start is the same as the probability of him being ruined if he acquires \$$n$ at any moment during the game. If at some moment during the game he has \$$n$, he can be ruined in two ways: by winning the next game and be ruined with \$$n + 1$, or by losing and then being ruined with \$$n - 1$. Thus

$$p_n = qp_{n+1} + (1 - q)p_{n-1}. \tag{2.7}$$

Replacing n by $n + 1$ and dividing by q, we obtain

$$p_{n+2} - \frac{1}{q}p_{n+1} + \frac{1 - q}{q}p_n = 0, \tag{2.8}$$

with $n = 0, 1 \ldots, N$. This is a second-order linear difference equation which requires two side conditions. While in the previous cases the initial conditions were a natural choice and thus we have not pondered on them, here the situation is slightly untypical. Namely, we know that the probability of ruin starting with \$0 is 1, hence $p_0 = 1$. Further, if the player has \$$N$, then he quits and cannot be ruined, so that $p_N = 0$. These are not initial conditions but an example of two-point conditions, that is, conditions prescribed at two arbitrary points. Such problems do not always have a solution.

To find the general solution of (2.8) we can use the method described in Section 1.1.2. The characteristic equation

$$\lambda^2 - q^{-1}\lambda + (1 - q)q^{-1} = 0$$

has roots $\lambda_1 = \frac{1-q}{q}$ and $\lambda_2 = 1$. Thus, according to the discussion in Section 1.1.2, if $q \neq 1/2$, then the general solution is

$$p_n = c_1 + c_2 \left(\frac{1-q}{q} \right)^n,$$

while if $q = 1/2$, then $\lambda_1 = \lambda_2 = 1$ and $p_n = c_1 + c_2 n$, where c_1 and c_2 are constants. To find the solution for the given boundary conditions, we denote $Q = (1-q)/q$ so that for $q \neq 1/2$

$$1 = c_1 + c_2, \qquad 0 = c_1 + Q^N c_2,$$

from where $c_2 = 1/(1 - Q^N)$, $c_1 = -Q^N/(1 - Q^N)$ and hence

$$p_n = \frac{Q^n - Q^N}{1 - Q^N}.$$

Analogous considerations for $q = 1/2$ yield $p_n = 1 - n/N$. For example, if $q = 1/2$ and the gambler starts with $n = \$20$ with the target $N = \$1000$, then $p(20) = 1 - 20/1000 = 0.98$; that is, his ruin is almost certain.

In general, if the gambler plays a long series of games, which can be modelled here as taking $N \to \infty$, then he will be ruined almost certainly even if the game is fair ($q = \frac{1}{2}$).

Exercise 2.4 You play the following game: on each play, the probability that you win \$2 is 0.1, the probability that you win \$1 is 0.3, and the probability that you lose 1 is 0.6. Suppose that you quit when either you are broke or when you have at least \$N. Write the third-order equation (together with the boundary conditions) that describes the probability p_n of eventually going broke if you have \$n.

2.2 Difference equations of population theory

The modelling process in the examples discussed above was relatively simple and only involved translation of the given rules into mathematical relations. This was due to the fact that there was no need to discover these rules as they were explicitly stated in the bank's or game's regulations. In this section we shall attempt to model behaviour of more complicated systems, when the modelling involves making some hypotheses about the rules governing them.

2.2.1 Single equations for unstructured population models

In many fields of human endeavour it is important to know how various populations grow and what factors influence their growth. Knowledge of this kind is important in studies of bacterial growth, wildlife management, ecology and harvesting.

Many animals tend to breed only during a short, well-defined, breeding season. If we neglect death, it is then natural to think of the population as only changing intermittently and to measure time discretely, using positive integers corresponding to the breeding seasons. Hence the obvious approach for describing the growth of such a population is to write down a suitable difference equation relating the size of the population in a given season to the size in the preceding ones. Later we shall also look at populations of species that breed continually, such as humans.

We begin with the simplest population models, discuss their advantages and drawbacks and build more realistic variants to address the latter.

2.2.2 Exponential growth – linear first-order difference equations

Let us start with semelparous populations; that is, ones characterized by a single reproductive episode before death. Typical examples of such populations are insects, who often have well-defined annual non-overlapping generations – adults lay eggs in spring/summer and then die. The eggs hatch into larvae which feed and grow and later undergo metamorphosis to spend winter in the so-called pupal stage. The adults, also called imago, emerge from the pupae in spring. We take the census of adults in the breeding seasons. Then it is natural to describe the population as the sequence of numbers

$$N_0, N_1, \ldots, N_k,$$

where N_k is the number of adults in the kth breeding season.

The simplest assumption to make is that there is a functional dependence between subsequent generations

$$N_{n+1} = f(N_n), \quad n = 0, 1, \ldots \tag{2.9}$$

Let us introduce the number R_0, which is the average number of eggs laid by an adult. We call R_0 the *intrinsic growth rate*. The simplest functional dependence (2.9) is

$$N_{n+1} = R_0 N_n, \quad n = 0, 1, \ldots, \tag{2.10}$$

which describes the situation in which the size of the population is determined only by its fertility. Note, that the fact that there are no surviving adults after breeding is implicit in the form of (2.10) and the definition of R_0.

The exponential (or Malthusian) equation (2.10) has a much larger range of applications than the one described above. For instance, in most populations the generations, in general, do overlap. To adapt (2.10) to such a situation, we will look at large populations, in which individuals give birth to new offspring but also die after some time. We treat such a population as a whole, assuming that its growth is governed by the average behaviour of its individual members. Thus, we make the following assumptions:

- each member of the population produces on average the same number of offspring;
- each member has an equal chance of dying (or surviving) before the next breeding season;
- the ratio of females to males remains the same in each breeding season.

We also assume:

- age differences between members of the population can be ignored;
- the population is isolated – there is no immigration or emigration.

Suppose that, on average, each member of the population gives birth to the same number of offspring, β, each season. The constant β is called the per capita *birth rate*. We also define μ as the probability that an individual will die before the next breeding season and call it the per capita *death rate*. Thus, denoting by N_k the number of individuals of the population at the start of the kth breeding season, we obtain $N_{k+1} = N_k - \mu N_k + \beta N_k$; that is

$$N_{k+1} = (1 + \beta - \mu)N_k = RN_k, \qquad (2.11)$$

where R is the *net growth rate*. This equation reduces to (2.10) by putting $\mu = 1$ (so that the whole adult population dies) and $\beta = R_0$. Mathematically, (2.10) and (2.11) are the same equation, the only difference being the interpretation of the growth parameter.

Equation (2.11) is easily solved, yielding

$$N_k = R^k N_0, \qquad k = 0, 1, 2 \ldots, \qquad (2.12)$$

with R replaced by R_0 in semelparous populations. We see that the behaviour of the model depends on R. If $R < 1$, then the population decreases towards extinction, but with $R > 1$ it grows indefinitely. Such behaviour over long periods of time is not observed in any population so it is clear that the model is over-simplified and requires corrections.

Exercise 2.5   A population of birds on an island has constant per capita birth (β) and death (μ) rates. Also a constant number I of birds migrate to the island each year. Find the difference equation describing the growth of birds' population on this island, assuming that the newcomers start laying eggs only one year after arrival on the island.

2.2.3 The death rate μ and the average lifespan of an individual

The death rate μ introduced above can be given another interpretation. Denote by $P(k)$ the probability that an individual, born at $k = 0$, is alive at time k; that is, at the end of the kth season. Now, in order to be alive at time k, the individual had to be alive at the end of the $(k-1)$th season and could not die between $k-1$ and k. Since the set of individuals alive at $k-1$ is a disjoint union of the sets of individuals alive at k and those who died between $k-1$ and k, the probability of dying between $k-1$ and k is

$P(k-1) - P(k)$, see e.g. (Feller, 1968). On the other hand, assuming that the probability μ of dying in a given season is constant in time, we see that that the probability that an individual dies at k is $\mu P(k-1)$. Indeed, to die at the end of the kth season, the individual had to be alive at time $k-1$ and then die. Comparing these two expressions, we arrive at

$$P(k) = (1-\mu)P(k-1), \qquad P(0) = 1;$$

that is

$$P(k) = (1-\mu)^k.$$

The average lifespan L means the expected duration of life. To find it, we observe that, by the above considerations, the probability of dying at age k is $\mu P(k-1) = \mu(1-\mu)^{k-1}$. Without using probability theory, it can be explained as follows: after first year a proportion μ of the population dies and $1-\mu$ survives, then after second year a proportion μ of them die; that is, a proportion $\mu(1-\mu)$ of the initial population lives just 2 years and $(1-\mu)^2$ lives on, etc. Thus, the average life span is found to be

$$L = \mu \sum_{k=1}^{\infty} k(1-\mu)^{k-1} = \frac{1}{\mu}, \qquad (2.13)$$

where we used

$$\sum_{k=1}^{\infty} k z^{k-1} = \sum_{k=1}^{\infty} \sum_{j=1}^{k} z^{k-1} = \sum_{j=1}^{\infty} \sum_{k=j}^{\infty} z^{k-1}$$

$$= \sum_{j=1}^{\infty} z^{j-1} \sum_{r=0}^{\infty} z^r = \frac{1}{(1-z)^2}$$

for $z = 1 - \mu$.

2.2.4 Models leading to nonlinear difference equations

In real populations, some of the R_0 offspring produced on average by each adult will not survive to be counted as adults in the next census. If we denote by $S(N)$ the *survival rate*; that is, the fraction of the population that survives over the season, then the Malthusian equation for semelparous populations, (2.10), is replaced by a more general equation which can be written in one of the following forms

$$N_{k+1} = R_0 S(N_k) N_k = F(N_k) N_k = f(N_k), \qquad (2.14)$$

$k = 0, 1, \ldots$, where $F(N)$ is the effective per capita reproduction rate when the population is of size N. Again we emphasise that here we describe semelparous populations; appropriate terms would have to be added to (2.14) to describe populations with adults surviving the breeding season, as in (2.18).

Such models typically lead to nonlinear equations. We introduce some typical nonlinear models by specifying the reproduction rates in (2.14).

Beverton–Holt type models. Let us look at the model (2.14). We would like it to display a *compensatory* behaviour; that is, the mortality should balance the increase in number of individuals. For this we should have $NS(N) \approx const.$ Also, for small N, $S(N)$ should be approximately 1 as we expect very small intra-species competition and therefore the growth should be exponential with the growth rate R_0. A simple function of this form is

$$S(N) = \frac{1}{1 + aN},$$

leading to

$$N_{k+1} = \frac{R_0 N_k}{1 + a N_k}.$$

Let us introduce the concept of the *carrying capacity* of the environment. This is the number K such that if the population reaches K, it will stay there or, in other words, if $N_k = K$ for some k, then $N_{k+m} = K$ for all $m \geq 0$. Substituting $N_{k+1} = N_k = K$ in the equation above gives

$$K(1 + aK) = R_0 K,$$

leading to $a = (R_0 - 1)/K$ and the resulting model, called the *Beverton–Holt model*, takes the form

$$N_{k+1} = \frac{R_0 N_k}{1 + \frac{R_0 - 1}{K} N_k}. \qquad (2.15)$$

As we said earlier, this model is compensatory.

A generalization of this model is called the *Hassell*, or again *Beverton–Holt*, model and reads

$$N_{k+1} = \frac{R_0 N_k}{(1 + aN_k)^b}. \qquad (2.16)$$

Substituting $x_k = aN_k$ reduces the number of parameters in (2.16), giving

$$x_{k+1} = \frac{R_0 x_k}{(1 + x_k)^b} \qquad (2.17)$$

which will be analysed in Section 4.2.5.

Remark 2.6 While the derivation of the Beverton–Holt equation presented above may seem to be ad hoc, we shall see in Sections 5.1.4 and 5.2.2 that it appears in a natural way as a discretization of well-established population models.

The logistic equation. The Beverton–Holt models are best applied to semelparous insect populations but have been also used in fisheries models. For populations surviving to the next cycle, it is more informative to write the difference equation in the form

$$N_{k+1} = N_k + F(N_k)N_k, \qquad (2.18)$$

so that the per capita increase in the population is given by $F(N) = R_0 S(N)$. Note that the first term on the right hand side of (2.18) was not present in (2.14) due to the fact that in semelparous populations we have no adult survivals from one season to another. Here, in contrast, we assume that no adults die (though death can be incorporated, e.g., by introducing a factor $d < 1$ in front of the first N_k).

As before, the function F can have different forms but should satisfy the following requirements:

- Due to overcrowding, $F(N)$ must decrease as N increases until N equals the carrying capacity K; then $F(K) = 0$ and, as above, $N = K$ stops changing.
- Since for N much smaller than K there is small intra-species competition, we should observe an exponential growth of the population so that $F(N) \approx R_0$ as $N \to 0$; here R_0 is called the unrestricted growth rate of the population.

The constants R_0 and K are usually determined experimentally.

In the spirit of mathematical modelling we start with the simplest function satisfying these requirements. This is a linear function. To satisfy the above conditions, it must be chosen as

$$F(N) = -\frac{R_0}{K}N + R_0.$$

Substituting this formula into (2.18) yields the so-called *discrete logistic equation*

$$N_{k+1} = N_k + R_0 N_k \left(1 - \frac{N_k}{K}\right), \tag{2.19}$$

which is still one of the most often used discrete equations of population dynamics.

By substitution

$$x_k = \frac{R_0}{1 + R_0}\frac{N_k}{K}, \qquad \mu = 1 + R_0,$$

we can reduce (2.19) to a simpler form

$$x_{k+1} = \mu x_k (1 - x_k), \tag{2.20}$$

which typically is used for mathematical analysis.

Exercise 2.7   Show that the logistic equation (2.19) can be derived from the general equation (2.11) by assuming that the mortality μ is not constant but equals

$$\mu = \mu_0 + \mu_1 N.$$

Here μ_0 corresponds to death by natural causes and μ_1 could be attributed to e.g. cannibalism, where one adult eats/kills on average a portion μ_1 of the population.

Exercise 2.8   Show that for N_k close to K in (2.19), subsequent iterations can produce negative values for the population. What can be an interpretation of negative values of N_k?

Exercise 2.9   Show that the equation

$$N(k+1) = N_k e^{r(1-\frac{N_k}{K})}, \tag{2.21}$$

called the Ricker equation, satisfies the assumption introduced in constructing the logistic equation but does not allow for negative solutions. What happens if N_k is (a) slightly smaller and (b) slightly larger than K?

Exercise 2.10 Another drawback of the previous models is that they describe situations in which the growth rate only decreases when the population increases. However, in 1931 Warder Clyde Allee, see e.g., (Thieme, 2003), noticed that in small, or dispersed, populations, individual chances of survival decrease which can lead to their extinction. This effect could be due to the difficulties in finding a mating partner or a more difficult cooperation in organizing defence against predators. Consider a population described by

$$N_{k+1} = N_k(1 + r(L - N_k)(N_k - K)), \qquad (2.22)$$

where $0 < L < K$. Show that $N_{k+1} < N_k$ if $0 < N_k < L$ and $N_{k+1} > N_k$ if $L < N_k < K$ so that (2.22) gives an example of the behaviour described by Allee.

2.3 Some applications of discrete population models

In this section we discuss some problems, the solution of which requires a good understanding of the model without necessarily solving the relevant equations. Thus, the approach presented here could serve as an introduction to the qualitative theory considered in Chapter 4.

2.3.1 Estimates of parameters in population models

Most population models contain parameters which are not given and must be determined by fitting the model to the observable data. We consider two problems of this type.

Growth rate in an exponential model. A total of 435 fish were introduced in 1979 and 1981 into a lake. In 1989, the commercial net catch alone was 4 000 000 kg. Since the growth of this population was so fast, it is reasonable to assume that it obeyed the Malthusian law $N_{k+1} = RN_k$. Assuming that the average weight of a fish is 1.5 kg, and that in 1999 only 10% of the fish population was caught, we seek lower and upper bounds for R.

The solution is based on a realization that we have two extreme cases: either all fish introduced in 1979 died or all fish introduced in 1981 died. For a given output, the former scenario would give the lowest growth rate while the highest would follow from the latter. To put this in mathematical terms, we recall equation (2.12). We have

$$N_k = N_0 R^k,$$

where we measure k in years and $R > 1$ (as we have growth). Let us denote by $N^{(1)}$ and $N^{(2)}$ the numbers of fish introduced in 1979 and 1981, respectively, so that $N^{(1)} + N^{(2)} = 435$. Thus, we can write the equation

$$N_{1989} = N^{(1)} R^{1989-1979} + N^{(2)} R^{1989-1981} = N^{(1)} R^{10} + N^{(2)} R^8.$$

Since we know neither $N^{(1)}$ nor $N^{(2)}$, we observe that $R^2 > 1$ and thus

$$N_{1989} \leq N^{(1)} R^{10} + N^{(2)} R^{10} = 435 R^{10}.$$

Similarly

$$N_{1989} \geq N^{(1)} R_0^8 + N^{(2)} R^8 = 435 R^8.$$

Hence

$$\sqrt[10]{\left(\frac{N_{1989}}{435}\right)} \leq R \leq \sqrt[8]{\left(\frac{N_{1989}}{435}\right)}.$$

Now, the data of the problem give $N_{1989} = 10 \times 4000000/1.5 \approx 2666666$ and so

$$2.39 \leq R \leq 2.97.$$

Exercise 2.11   A population grows according to the logistic equation

$$N_{k+1} = N_k + R_0 N_k \left(1 - \frac{N_k}{K}\right),$$

with R_0 and K unknown. Show that to determine R_0 and K, it is enough to know three subsequent measurements of the population size, when it is not in equilibrium.

2.3.2 Sustainable harvesting

Let us consider a pond in which the number of fish grows according to the logistic equation (2.19),

$$N_{k+1} = N_k + R_0 N_k \left(1 - \frac{N_k}{K}\right).$$

This equation only can be solved in some particular cases, see Section 2.4. However, even without solving it, we can draw a conclusion of some importance for fisheries.

The basic idea of a sustainable economy is to find an optimal level of fishing: too much harvesting would deplete the fish population beyond recovery and too little would provide insufficient return for the community. To maintain the population at a constant level, only the increase in the population should be harvested during any one season. In other words, the harvest should be $H_k = N_{k+1} - N_k$. Using the logistic equation, we find

$$H_k = R_0 N_k \left(1 - N_k/K\right).$$

Hence, to maximize the yield at each k, the population should be kept at the size $N_k = N$ for which the right hand side attains the absolute maximum. We note that the right hand side is a quadratic function, $f(N) = R_0 N \left(1 - N/K\right)$, and it is easy to find that the maximum is attained at $N = K/2$; that is, the population should be kept at around half of the carrying capacity of the environment. Thus, the maximum sustainable yield is

$$H = R_0 K/4. \tag{2.23}$$

Exercise 2.12   A more realistic version of the above problem can be derived by introducing actual fishing in the model. Then we obtain

$$N_{k+1} = N_k + R_0 N_k \left(1 - \frac{N_k}{K}\right) - qEN_k,$$

where E is the fishing effort (e.g., the number of fishing vessels in the area) and q is the fishing efficiency; that is, the fraction of the total fish population caught by one vessel in the season. Using an argument as above,

(i) show that fish stock can be kept at a constant size with fishing only if $qE < R_0$ and find the size of the population in this case;

(ii) find the fishing yield for such a population;

(iii) given that q, R_0 and K are constant, show that the maximum sustainable yield of $R_0 K/4$ (compare with (2.23)) is attained for the fishing effort $E = R_0/2q$;

(iv) show that if $E > R_0/2q$, then the yield decreases with growing E.

Using (iv), draw the conclusion that greed does not pay.

2.4 Some explicitly solvable nonlinear models

We complete this chapter by presenting some nonlinear models which can be explicitly solved by appropriate substitutions.

2.4.1 The Beverton–Holt model

We recall that the Beverton–Holt equation, (2.16), can be simplified to

$$x_{n+1} = \frac{R_0 x_n}{(1 + x_n)^b}. \tag{2.24}$$

While for general b this equation can display very rich dynamics, which will be looked at in Section 4.2.5, for $b = 1$ it can be solved explicitly. So, let us consider

$$x_{n+1} = \frac{R_0 x_n}{1 + x_n}. \tag{2.25}$$

We can recognize this equation as an equation of the Ricatti type (1.9) so that the substitution $y_n = 1/x_n$ converts (2.25) into

$$y_{n+1} = \frac{1}{R_0} + \frac{1}{R_0} y_n.$$

Using (1.5), we find

$$y_n = \frac{1}{R_0} \frac{R_0^{-n} - 1}{R_0^{-1} - 1} + R_0^{-n} y_0 = \frac{1 - R_0^n}{R_0^n (1 - R_0)} + R_0^{-n} y_0$$

if $R_0 \neq 1$ and

$$y_n = n + y_0$$

for $R_0 = 1$. From these equations, we see that $x_n \to R_0 - 1$ if $R_0 > 1$ and $x_n \to 0$ if $R_0 \leq 1$ as $n \to \infty$. It is maybe surprising that a population faces extinction if $R_0 = 1$ (which corresponds to every individual giving birth to one offspring on average). However, the density depending factor causes some individuals to die between reproductive seasons which means that the population with $R_0 = 1$ in fact decreases with every cycle.

2.4.2 The logistic equation

In general, the discrete logistic equation does not admit closed form solutions and also displays a very rich dynamics, see Section 4.2.4. However, some special cases can be solved by an appropriate substitution. We will look at two such cases. First consider

$$x_{n+1} = 2x_n(1 - x_n) \qquad (2.26)$$

and use the substitution $x_n = 1/2 - y_n$. Then

$$\frac{1}{2} - y_{n+1} = 2 \left(\frac{1}{2} - y_n \right) \left(\frac{1}{2} + y_n \right) = \frac{1}{2} - 2y_n^2,$$

so that $y_{n+1} = 2y_n^2$. We see that if $y_0 = 0$, then $y_n = 0$ for all $n \geq 1$. Furthermore, the solution y_n for $n \geq 1$ does not change if we change the sign of y_0. Thus, we can take $|y_0| > 0$ as the initial condition. Then $y_n > 0$ for $n \geq 1$ and we can take the logarithm of both sides getting, for $n \geq 1$, $\ln y(n+1) = 2 \ln y_n + \ln 2$ which, upon substituting $z_n = \ln y_n$, becomes the inhomogeneous linear equation $z_{n+1} = 2z_n + \ln 2$. Using (1.5), we find the solution to be $z_n = 2^n z_0 + \ln 2(2^n - 1)$. Hence

$$y_n = e^{z_n} = e^{2^n \ln |y_0|} e^{\ln 2(2^n - 1)} = y_0^{2^n} 2^{2^n - 1},$$

where we dropped the absolute value bars as we raise y_0 to even powers. Thus

$$x_n = \frac{1}{2} - \left(\frac{1}{2} - x_0 \right)^{2^n} 2^{2^n - 1}.$$

We note that for $x_0 = 1/2$ we have $x_n = 1/2$ for all n, so that we obtain a constant solution. In other words, $x = 1/2$ is an equilibrium point of (2.26), see Section 4.2.

Another particular logistic equation which can be solved by substitution is

$$x_{n+1} = 4x_n(1 - x_n). \qquad (2.27)$$

First we note that since $f(x) = 4x(1 - x) \leq 1$ for $0 \leq x \leq 1$, we have $0 \leq x_{n+1} \leq 1$ if x_n has this property. Thus, assuming $0 \leq x_0 \leq 1$, we can use the substitution

$$x_n = \sin^2 y_n \qquad (2.28)$$

which yields

$$x_{n+1} = \sin^2 y_{n+1} = 4\sin^2 y_n(1 - \sin^2 y_n)$$
$$= 4\sin^2 y_n \cos^2 y_n = \sin^2 2y_n.$$

This gives the family of solutions

$$y_{n+1} = \pm 2y_n + k\pi, \qquad k \in \mathbb{Z}.$$

However, bearing in mind that our aim is to find x_n given by (2.28) and using the periodicity and symmetry of the function $\sin^2$, we can discard $k\pi$ as well as the minus sign and focus on $y_{n+1} = 2y_n$. This is a geometric progression and we get $y_n = C2^n$, where $C \in \mathbb{R}$ is arbitrary, as the general solution. Hence

$$x_n = \sin^2 C2^n,$$

where C is to be determined from $x_0 = \sin^2 C$. What is remarkable in this example is that, despite the fact that there is an explicit formula for the solution, the dynamics generated by (2.27) is very irregular (chaotic), see Section 4.2.

3

Basic differential equations models

In the previous section we saw that difference equations can be used to model quite a diverse phenomena but their applicability is limited by the fact that the system should not change between subsequent time steps. These steps can vary from fractions of a second to years or centuries but they must stay fixed in the model. On the other hand, there are numerous situations when changes can occur at all times. These include the growth of populations in which breeding is not restricted to specific seasons, motion of objects, where the velocity and acceleration may change at every instant, spread of an epidemic with no restriction on infection times, and many others. In such cases it is not feasible to model the process by relating the state of the system at a particular instant to a finite number of earlier states (although this part remains as an intermediate stage of the modelling process). Instead, we have to find relations between the rates of change of quantities relevant to the process. The rates of change typically are expressed as derivatives and thus continuous time modelling leads to differential equations which involve the derivatives of the function describing the state of the system.

In what follows we shall derive several differential equation models trying to provide continuous counterparts of some discrete systems described above.

3.1 Equations related to financial mathematics

In this section we shall provide continuous counterparts of equations (2.2) and (2.5) and compare the results.

3.1.1 Continuously compounded interest and loan repayment

Many banks advertise continuous compounding of interest. This means that the conversion period α of Section 2.1 tends to zero so that the interest is added to the account on a continual basis. Let us measure time in years, let Δt be the conversion period and p the annual interest rate. Then the increase in the deposit between time instants t and $t + \Delta t$ is

$$S(t + \Delta t) = S(t) + \Delta t \frac{p}{100} S(t). \qquad (3.1)$$

Dividing this relation by Δt and passing with Δt to zero, as suggested by the definition of continuously compounded interest, yields the differential equation

$$\frac{dS}{dt} = \bar{p}S, \qquad (3.2)$$

where $\bar{p} = p/100$. We note that here and elsewhere in this book we assume that the function describing the state of the system (such as $S(t)$ in this model) is differentiable so that the passage to the derivative is justified. It is important to realize that, in general, this property is far from obvious and sometimes we have to work with discrete equations even in continuous models.

Equation (3.2) is a first order linear equation. It is easy to check that it has the solution

$$S(t) = S_0 e^{\bar{p}t}, \qquad (3.3)$$

where S_0 is the initial deposit made at time $t = 0$.

To compare this formula with the discrete one (2.2), we note that in t years we have $k = t/\alpha$ conversion periods, so that

$$S(t) = N_k = (1 + \bar{p}\alpha)^k S_0 = (1 + \bar{p}\alpha)^{t/\alpha} S_0 = \left((1 + \bar{p}\alpha)^{1/\bar{p}\alpha}\right)^{\bar{p}t}.$$

From calculus we know that

$$\lim_{x \to 0^+} (1+x)^{1/x} = e,$$

and that the function $x \to (1+x)^{1/x}$ is monotonically increasing. Thus, if the interest is compounded very often (almost continuously), then practically

$$S(t) \approx S_0 e^{\bar{p}t}, \qquad (3.4)$$

which is exactly (3.3). It is clear that after 1 year the initial investment will increase by the factor $e^{\bar{p}}$ and, recalling (2.4), we have the identity

$$1 + p_{\text{eff}} = e^{\bar{p}}, \qquad (3.5)$$

which can serve as the definition of the effective interest rate when the interest is compounded continuously. This relation can, of course, be obtained by passing with α to 0 in (2.4). Typically, contrary to (2.2), the exponential can be calculated even on a simple calculator, which makes (3.4) easier to use. Of course, it does not give the exact solution for any relevant discrete equation but, due to the monotonicity of the limit, the continuously compounded interest rate is the best one can get and the difference between the result obtained by the continuous and the discrete formulae is negligible. A short calculation reveals that if one invests \$10000 at $p = 15\%$ in banks with conversion periods of 1 year, 1 day and with continuously compounded interest, then the return will be, respectively, \$11500, \$11618 and \$11618.3. That is why the continuous formula (3.3) can be used as a good approximation for the real returns with small, but non-zero, conversion periods.

3.1.2 Continuous loan repayment

Let us assume that an annual interest $p\%$ is being paid off continuously at a rate of $\rho > 0$ per annum. Applying to the discrete equation (2.5) an argument similar to the one used in the previous subsection, we find that the equation governing the continuous loan repayment is

$$\frac{dD}{dt} - \bar{p}D = -\rho, \qquad (3.6)$$

where $D(t)$ is the outstanding debt at time t, where $\bar{p} = p/100$. Assume further that the original amount borrowed from a bank is $D(0) = D_0$. This is a Cauchy problem for an inhomogeneous linear equation, discussed in Section 1.3.2. The integrating factor is $\exp(-\bar{p}t)$ and hence we can write the equation as

$$(De^{-\bar{p}t})' = -\rho e^{-\bar{p}t}$$

which, upon integration from 0 to t and using the initial condition, yields

$$D(t) = D_0 e^{\bar{p}t} + \frac{\rho}{\bar{p}}(1 - e^{\bar{p}t}).$$

As in Section 2.1.2, let us determine the instalments for a given loan D_0, annual interest rate p and repayment period T. Setting $D(T) = 0$, we get

$$\rho = \frac{\bar{p}D_0 e^{\bar{p}T}}{e^{\bar{p}T} - 1}. \qquad (3.7)$$

Let us test this formula against the numerical data used in Section 2.1.2. We have $D_0 = \$200000$, $\bar{p} = 0.13$ and $T = 20$ (remember that we use years, and not the conversion period as in the discrete case, as units of time). Then we get $\rho = 28086.1$. This is, however, the annual instalment, thus the amount paid per month is $R = \$2340.5$. This gives a slightly better rate than if instalments were paid on a monthly basis but still (3.7) can be used as a good approximation of the real instalments.

Indeed, taking in the discrete formula (2.6) the annual payment $\rho_\alpha = R/\alpha$ and $n = T/\alpha$, where T is time in years and $1/\alpha$ is the number of payments in a year, we get

$$\lim_{\alpha \to 0} \rho_\alpha = \lim_{\alpha \to 0} \frac{\bar{p}D_0}{(1 - (1 + \alpha\bar{p})^{1/\alpha\bar{p}})^{-T\bar{p}}} = \frac{\bar{p}D_0}{1 - e^{-\bar{p}T}},$$

which, after simple algebra, becomes (3.7). Moreover, the limit is monotonic, so that indeed $\rho_\alpha < \rho$ for any $\alpha > 0$.

3.2 Radiocarbon dating

Exponential growth models appear in numerous processes, when-
ever the rate of change of some quantity is proportional to the
amount present. One of them is radioactive decay.

Radioactive substances undergo a spontaneous decay due to the
emission of α particles. The mass of α particles is small in comparison with the mass of the sample of the radioactive material so it
is reasonable to assume that the function N, giving the number of
particles in the sample, is continuous. Experiments indicate that
the rate of decrease of the mass of the sample is proportional to
its mass. This principle immediately leads to the equation

$$N' = -kN, \tag{3.8}$$

where N is the number of radioactive particles present in the
sample and k is a proportionality constant.

One of the best known applications of this model is the ra-
diocarbon dating used for finding the age of samples which once
contained a living matter, like fossils, etc. It is based on the observation that the element carbon appears in nature as a mixture
of stable Carbon-12 (C^{12}) and radioactive Carbon-14 (C^{14}) and
the ratio between them has remained constant throughout his-
tory. Thus, while an animal or a plant is alive, the ratio C^{14}/C^{12}
in its tissues is a known constant. When it dies, however, there
is no new carbon absorbed by the tissues and, since the radioactive C^{14} decays, the ratio C^{14}/C^{12} decreases as the amount of C^{14}
decreases.

To be able to use (3.8) we note that, similarly to (3.3), its
solution is given by

$$N(t) = N(t_0)e^{-k(t-t_0)}. \tag{3.9}$$

Then we we must find a way to determine the value of k for C^{14}.
What can be directly measured is the number of particles which
remain in the sample after some time (through the mass of the
sample). The parameter which is most often used when dealing
with radioactive materials is the so-called *half-life* defined as the
time T after which only half of the original number of particles
remains. This is a constant, depending only on the material and

not on the original number of particles or the moment in time at which we started to observe the sample. The relation between k and T can be found from the equation

$$N(T + t_0) = 0.5N(t_0) = N(t_0)e^{-kT};$$

that is, $kT = \ln 2$ so that the solution is given by

$$N(t) = N(t_0)e^{-(t-t_0)\ln 2/T}. \tag{3.10}$$

It is clear that after time T, the number of radioactive particles in a sample halves, irrespective of the initial amount and the initial time t_0, so that indeed the number of particles halves after every period of length T.

To demonstrate how this formula is applied in a concrete case, we estimate the age of a charcoal sample found in 2003 in a prehistoric cave, in which the ratio C^{14}/C^{12} was 14.5% of its original value. The half-life of C^{14} is 5730 years.

The crucial step here is to translate the absolute numbers of C^{14} appearing in (3.10) into the ratio C^{14}/C^{12} which is the only information available from the experiment. Imagine a reference sample containing a certain amount $N_{14}(t_0)$ of C^{14} and $N_{12}(t_0)$ of C^{12} at some time t_0. Then, at time t, we will have $N_{14}(t)$ of C^{14} but for C^{12} the amount does not change: $N_{12}(t) = N_{12}(t_0)$. Thus, dividing

$$N_{14}(2003) = N_{14}(t_0)e^{-(2003-t_0)\ln 2/5730}$$

by the constant N_{12}, we will get the equation governing the evolution of the ratio C^{14}/C^{12}

$$0.145\frac{N_{14}(t_0)}{N_{12}(t_0)} = \frac{N_{14}(2003)}{N_{12}(2003)} = \frac{N_{14}(t_0)}{N_{12}(t_0)}e^{-(2003-t_0)\ln 2/5730}.$$

Thus $0.145 = e^{-(2003-t_0)\ln 2/5730}$; that is,

$$t_0 = 2003 + \frac{5730\ln 0.145}{\ln 2} = -13960$$

so that the sample dates back to around 12000 BCE.

3.3 Differential equations for population models

In this section we will study first-order differential equations which appear in population theory. At first glance it appears that it is impossible to model the growth of a species by differential equations since the population of any species always changes by integer amounts. Hence the population of any species can never be a differentiable function of time. However, similarly to the argument in the previous section, if the population is large and it increases by one, then the change is very small when compared to the size of this population. Thus often we may assume that large populations change continuously (and even in a differentiable way) in time and, if the final answer is not an integer, we shall round it to the nearest integer.

Another way of accepting non-integer entries in population models is by considering – not the whole population – but its density; that is, the number if individuals present in a unit area or volume of space. As so defined, the density of a population typically is not an integer.

A similar justification applies to our use of t as a real variable: in the absence of specific breeding seasons, reproduction can occur at any time and for sufficiently large populations it is natural to think of reproduction as occurring continuously.

Let $N(t)$ denote the size of a population of a given isolated species at time t and let Δt be a small time interval. Then the population at time $t + \Delta t$ can be expressed as

$$N(t + \Delta t) - N(t) = \text{number of births in } \Delta t$$
$$- \text{ number of deaths in } \Delta t.$$

It is reasonable to assume that the numbers of births and deaths in a short time interval is proportional to the population at the beginning of this interval and proportional to the length of this interval, thus

$$N(t+\Delta t) - N(t) = \beta(t, N(t))N(t)\Delta t - \mu(t, N(t))N(t)\Delta t. \quad (3.11)$$

Taking $r(t, N)$ to be the difference between the birth and death rate coefficients at time t for the population of size N, we obtain

$$N(t + \Delta t) - N(t) = r(t, N(t))\Delta t N(t).$$

Note that this step is the same as in the discrete time modelling with Δt being the unit of time. We are, however, interested in changes occurring over time intervals of arbitrary length. Thus, dividing by Δt and taking the limit $\Delta t \to 0$, we obtain the differential equation

$$N' = r(t, N)N. \tag{3.12}$$

Here both the coefficient $r(t, N)$ and N are unknown. To solve the equation we need to make some assumptions on r, which are guided by the experimental data. We will discuss several typical cases now.

3.3.1 Exponential growth

The simplest possible $r(t, N)$ is a constant and in fact such a model is used in short-term population forecasts. So, let us assume that $r(t, N(t)) = r$ which gives

$$N' = rN. \tag{3.13}$$

This is exactly the same equation as (3.2), (3.8) or (5.17) and thus we can use the same mathematical techniques as for the former models. As in (3.9), the solution to it is given by

$$N(t) = N(t_0)e^{r(t-t_0)}, \tag{3.14}$$

where $N(t_0)$ is the size of the population at some fixed time t_0.

To be able to give some numerical illustration for this equation, we need the coefficient r and the population at some time t_0. Following (Braun, 1983), we use the data of the U.S. Department of Commerce: it was estimated that the Earth's population in 1965 was 3.34 billion and that the population was increasing at an average rate of 2% per year during the decade 1960–1970. Thus $N(t_0) = N(1965) = 3.34 \times 10^9$ with $r = 0.02$, and (3.14) takes the form

$$N(t) = 3.34 \times 10^9 e^{0.02(t-1965)}. \tag{3.15}$$

To test the accuracy of this formula, let us calculate when the population of the Earth is expected to double. To do this, we solve the equation

$$N(T + t_0) = 2N(t_0) = N(t_0)e^{0.02T}.$$

Hence $2 = e^{0.02T}$ and $T = 50\ln 2 \approx 34.6$ years. This is in an excellent agreement with the present observed value of the Earth's

Year	Population in billions
1500	0.5
1804	1
1927	2
1960	3.335
1965	4
1999	6
2010	6.972

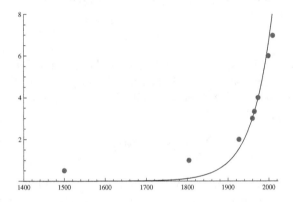

Figure 3.1 Comparison of actual population figures (points) with those obtained from (3.15)

population and also gives a good agreement with the observed data, provided in table below, if we do not go too far into the past, see Fig. 3.1.

On the other hand, if we try to extrapolate this model into the distant future, then we see that, say, in the year 2515, the population will reach 199980 ≈ 200000 billion. To realize what this means, recall that the Earth's total surface area is 510072000 square kilometres, 70.8% of which is covered by water; thus we have only 148940000 billion square kilometres to our disposal (less if water level rises!) and there will be only $0.7447\,\mathrm{m}^2$ ($= 86.3\,\mathrm{cm}\times 86.3\,\mathrm{cm}$)

per person. We can only hope that this model is not valid for all times.

3.3.2 The death rate and the average lifespan

Here we will show that the average lifespan is given by the same formula as in the discrete case, obtained in Section 2.2.3. Let μ be the death rate; that is, the probability that an individual dies in a time interval Δt is approximately equal to $\mu \Delta t$. Let $p(t)$ be the probability that the individual is alive at time t. Then the probability $p(t + \Delta t)$ of being alive at $t + \Delta t$, provided he was alive at t, is $p(t + \Delta t) = (1 - \mu \Delta t)p(t)$ which, as above, yields the differential equation

$$p' = -\mu p,$$

with $p(0) = 1$ (expressing the fact that the individual was born, and thus alive, at $t = 0$.) Hence we obtain $p(t) = e^{-\mu t}$. Arguing as in Section 2.2.3, we see that the average life span can be approximated by

$$\sum_{j=1} s_j(p(t_j) - p(t_{j+1})),$$

where we took an arbitrary partition of the maximum life span (here $[0, \infty)$), $0 = t_0 < t_1 < \cdots$ with $s_j \in [t_j, t_{j+1}]$. Then, using the definition of the Riemann integral, (Courant and John, 1999), and the differentiability of p we find, as in (2.13),

$$L = \int_0^\infty s \frac{d}{ds} e^{-\mu s} ds = -\mu \int_0^\infty s e^{-\mu s} ds = \frac{1}{\mu}. \qquad (3.16)$$

3.3.3 Logistic differential equation

As for discrete models, it has been observed that the exponential model for the population growth is satisfactory as long as the population is not too large. When the population gets very large (with regard to its habitat), these models cannot be very accurate, since they don't reflect the fact that the individual members have to compete with each other for limited living space, resources and food and then birth rate and death rate will depend on other variables. It is reasonable to assume that a given habitat can sustain only a finite number K of individuals, and the closer the population is to this number, the slower is its growth. As in the discrete case (2.19), the simplest way to take this into account is to employ the linear function $r(t, N) = R_0(K - N)$, where R_0 is the unrestricted growth rate. This choice yields the so-called *continuous logistic model*

$$N' = R_0 N \left(1 - \frac{N}{K} \right), \tag{3.17}$$

which has proved to be one of the most successful models for describing a single species population.

Exercise 3.1 Logistic equations appear in various contexts, see e.g. (Braun, 1983). Let us suppose that we have a community of constant size C and $N(t)$ members of this community have some important information at a time t. Assuming that

- the information is passed on when a person knowing it meets a person that does not,
- the rate at which one person meets other people is a constant f,

show that

$$N' = f N \left(1 - \frac{N}{C} \right);$$

that is, that the rate at which an information spreads in a closed community is governed by a logistic equation.

Exercise 3.2 Show that, by taking in (3.11) a constant birth rate β but a density-dependent mortality rate $\mu(N) = \mu_0 + \mu_1 N$, one obtains the logistic equation (3.17).

In general, show that taking in (3.11) a constant β and $\mu(N) = \mu_0 + \mu_1 N^\theta$, $\theta > 0$, yields the equation

$$N' = (\beta - \mu_0)N - \mu_1 N^{\theta+1}, \tag{3.18}$$

which is an equation of Bernoulli type.

Next we consider the Cauchy problem for (3.17),

$$N' = R_0 N \left(1 - \frac{N}{K} \right),$$

$$N(t_0) = N_0. \tag{3.19}$$

This is an example of a separable equation. Using the approach described in Section 1.3.1, we separate the variables to obtain

$$\frac{K}{R_0} \int_{N_0}^{N} \frac{ds}{(K - s)s} = t - t_0.$$

To integrate the left hand side, we use partial fractions

$$\frac{1}{(K - s)s} = \frac{1}{K} \left(\frac{1}{s} + \frac{1}{K - s} \right),$$

so that

$$t - t_0 = \frac{K}{R_0} \int_{N_0}^{N} \frac{ds}{(K - s)s} = \frac{1}{R_0} \ln \frac{N}{N_0} \left| \frac{K - N_0}{K - N} \right|.$$

Since $N(t) \equiv K$ and $N(t) \equiv 0$ are solutions of the logistic equation, the Picard theorem ensures that $N(t)$ cannot reach K in any finite time so, if $N_0 < K$, then $N(t) < K$ for any t, and if $N_0 > K$, then $N(t) > K$ for all $t > 0$. Therefore $(K - N_0)/(K - N(t))$ is always positive and, exponentiating, we get

$$e^{R_0(t - t_0)} = \frac{N(t)}{N_0} \frac{K - N_0}{K - N(t)},$$

which, solved with respect to $N(t)$, yields

$$N(t) = \frac{N_0 K}{N_0 + (K - N_0)e^{-R_0(t - t_0)}}. \tag{3.20}$$

Let us examine (3.20) to see what kind of population behaviour it predicts. First observe that we have

$$\lim_{t \to \infty} N(t) = K,$$

hence our model correctly reflects the initial assumption that K is the maximal capacity of the habitat. Next,

$$N' = \frac{R_0 N_0 K (K - N_0)e^{-R_0(t - t_0)}}{(N_0 + (K - N_0)e^{-R_0(t - t_0)})^2} :$$

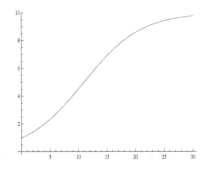

Figure 3.2 Logistic curve with $N_0 = 1, K = 10$ and $R_0 = 0.02$

thus, if $N_0 < K$, the population monotonically increases, whereas if we start with a population which is larger than the capacity of the habitat, then such a population will decrease towards K. Also

$$N'' = R_0(N(K - N))' = N'(K - 2N) = N(K - N)(K - 2N)$$

from which it follows that if we start from $N_0 < K$, then the population curve is convex down for $N < K/2$ and convex up for $N > K/2$. Thus, as long as the population is small (less then half of the capacity K), then the rate of growth increases, whereas for larger populations the rate of growth decreases. This results in the famous *logistic* or *S-shaped* curve which is presented in Fig. 3.2 for particular values of parameters $R_0 = 0.02, K = 10$ and $t_0 = 0$; that is, as the graph of

$$N(t) = \frac{10N_0}{N_0 + (10 - N_0)e^{-0.2t}}.$$

To show how this curve compares with the real data and with the exponential growth, we take (3.20) with the coefficients which, for the human population, are experimentally estimated as $K = 10.76$ billion and $R_0 = 0.029$. Then the logistic equation for the growth of the Earth's population will read

$$N(t) = \frac{N_0(10.76 \times 10^9)}{N_0 + ((10.76 \times 10^9) - N_0)e^{-0.029(t-t_0)}}.$$

We use this function with the value $N_0 = 3.34 \times 10^9$ at $t_0 = 1965$. The comparison is shown on Fig. 3.3.

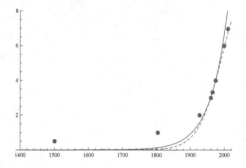

Figure 3.3 Human population on Earth. Comparison of observational data (points), exponential growth (solid line) and logistic growth (dashed line)

3.4 Equations of motion: second-order equations

Second-order differential equations appear often as equations of motion. This is due to Newton's law of motion that relates the acceleration of a body – that is, the second derivative of its position y with respect to time t – to the (constant) body mass m and the forces F acting on it:

$$y'' = F/m. \qquad (3.21)$$

We confine ourselves to a scalar, one-dimensional case with time-independent mass. The modelling in such cases concerns the form of the force acting on the body. We shall consider two such cases in detail.

3.4.1 A waste disposal problem

Toxic or radioactive waste is often disposed of by placing it in sealed containers that are then dumped at sea or deep lakes. Of concern is that these containers could crack upon hitting the sea or lake bed. Experiments confirmed that the drums can indeed crack if the velocity exceeds 4 m/s at the moment of impact. To make this procedure feasible, one has to find out the velocity of the container when it reaches the bed. Since typically the disposal takes place in deep waters, direct measurement is rather expensive and thus the problem should be solved by mathematical modelling.

As the container descends through the water, it is acted upon by three forces W, B, D. The weight W is given by $W = mg$, where g is the gravitational acceleration and m is the mass of the container. The buoyancy force B is the force of displaced water acting on the drum and its magnitude is equal to the weight of the displaced water; that is, $B = g\rho V$, where ρ is the density of the water and V is the volume of the container. If the density of the waste (together with its packaging) is smaller than the density of the water then, of course, the container will float. Thus we assume that the container is heavier than the displaced water and therefore it will start sinking. We can easily observe that any object moving through a medium like water, air, etc., experiences some resistance, called *drag*. The drag force always acts in the opposite direction to the motion of the object and its magnitude increases with increasing velocity. If the object moves very slowly in a viscous fluid, drag is proportional to the speed v of the moving object, otherwise the drag force is proportional to the square of the speed v and is magnitude is

$$D = cv^2 := \frac{c_d \rho A}{2} v^2, \tag{3.22}$$

where ρ is the density of the fluid, A is the cross-sectional area of the object perpendicular to the direction of the motion and c_d is the drag coefficient which depends on the shape of the body and is determined experimentally. If we now set $y = 0$ at the sea level and let the direction of increasing y be downwards, then from (3.21)

$$y'' = m^{-1} \left(W - B - c(y')^2 \right). \tag{3.23}$$

Since the container is simply dumped into the sea at the sea level, we supplement (3.23) with the initial conditions

$$y(0) = 0, \qquad y'(0) = v(0) = 0.$$

Equation (3.23) is a nonlinear second-order equations but it is reducible, see Section 1.3.4. Thus, we express the speed $v(t) = y'(t)$ as a function $\nu(y)$ of the depth y via

$$v(t) = \nu(y(t)).$$

By the chain rule, as in (1.26), we find

$$y'' = \frac{dv}{dt} = \frac{dv}{dy}\frac{dy}{dt} = v\frac{dv}{dy}.$$

Substituting this formula into (3.23), we obtain

$$mv\frac{dv}{dy} = \left(mg - B - cv^2\right). \qquad (3.24)$$

We have to supplement this equation with an appropriate initial condition. For this we have

$$\nu(0) = \nu(y(0)) = v(0) = 0.$$

Equation (3.24) is a separable equation which we can solve explicitly. First, we note that since

$$\pm\nu_T = \pm\sqrt{\frac{mg - B}{c}}$$

are (stationary) solutions to (3.24), by Picard's theorem, any solution with initial value in $(-\eta_T, \eta_T)$ is increasing and stays in this interval. Thus, $\nu(0) = 0$ implies $mg - B - cv^2(t) > 0$ for all t. Therefore we can divide both sides of (3.24) by $mg - B - cv^2$ and, integrating, we get

$$\int_0^\nu \frac{r\,dr}{mg - B - cr^2} = \frac{1}{m}\int_0^y ds = \frac{y}{m}.$$

Using $z = mg - B - cr^2$ with $dz = -2cr\,dr$, we obtain

$$\int \frac{r\,dr}{mg - B - cr^2} = -\frac{1}{2c}\int \frac{dz}{z} = -\frac{1}{2c}\ln(mg - B - cr^2) + C,$$

so that

$$-\frac{2cy}{m} = \ln\frac{mg - B - cv^2}{mg - B}.$$

Exponentiating and with some algebra we obtain

$$\nu(y) = \sqrt{\frac{mg - B}{c}}\left(1 - e^{-\frac{2c}{m}y}\right), \qquad (3.25)$$

where we used the fact that the speed is positive. We observe that,

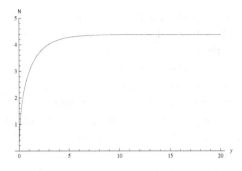

Figure 3.4 The speed of the drum as a function of its depth

as expected, the speed monotonically increases to a limit value, which is called the terminal velocity,

$$\nu_T = \sqrt{\frac{mg - B}{c}}.$$

It is the largest stationary solution to (3.24).

To answer the original question, we need concrete data. The containers are dropped into the 100 m deep sea. The dimensions of the cylinder are $h = 1\,\text{m}$ and $r = 0.25\,\text{m}$ so $V = 0.196\,\text{m}^3$. The average mass of the container and its contents is 500 kg. The density of the sea water is taken to be $1021\,\text{kg/m}^3$. Thus, the mass of the water displaced by the container is 200.5 kg. The calculation of the drag force is more complicated. Experiments show that the drag coefficient c_d in (3.22) for a flat bottomed cylinder may vary between 0.6 (if the cylinder sinks horizontally) to over 1, if it is released in the vertical position, see e.g., (Holland et al., 2004). Since we are interested in the worst case scenario, we take $c_d = 0.6$. In this case, the cross-sectional area is $2rh = 0.5\text{m}^2$ and hence

$$c = \frac{0.6 \cdot 0.5 \cdot 1021}{2} = 153.15.$$

Thus, we obtain $\nu_T = 4.38\,\text{m/s}$. This means that, though the speed of the container cannot be larger than 4.38 m/s, it will eventually exceed the safe speed of 4 m/s. By direct calculations, we find that this speed is already attained at $y = 2.931\,\text{m}$ (see Fig. 3.4) and hence the container could crack on impact.

Is it possible to avoid the danger of the containers cracking upon hitting the sea bed? One possibility is to use reinforced containers which could withstand higher impact velocities without cracking. However, this solution could be not feasible. Another option is to change the parameters of the problem, so that the terminal velocity becomes smaller than $4\,\mathrm{m/s}$. Looking at the data of the problem, we see that, given the container's properties and the location, the only parameter which we can control is the mass of the waste filling the container. Clearly, reduction of this mass will decrease the resultant force acting on the container, thus decreasing the terminal velocity, and it can be easily achieved by diluting the waste with, e.g., water which is lighter than the waste.

Exercise 3.3 Show that if the mass of the container together with the waste is smaller than $450\,\mathrm{kg}$ and the other parameters are left unchanged, then the terminal velocity ν_T is smaller than $4\,\mathrm{m/s}$.

Exercise 3.4 Find the speed of a container of waste as a function of the depth y if the drag force is zero $(c = 0)$. Since the velocity $\nu_0(y)$ of the drum without the drag is not smaller than the velocity $\nu(y)$ with a drag, show that the container will not crack upon impact if they are dropped into L metres of water with $(2\mathrm{g}(mg - B)L/mg)^{1/2} < 4$.

3.4.2 Motion in a changing gravitational field

According to Newton's law of gravitation, two objects of masses m and M attract each other with a force of magnitude $F = GmM/d^2$, where G is the gravitational constant and d is the distance between the objects' centres. Since at any planet's surface the force is equal (by definition) to $F = mg$, where g is the gravitational acceleration at the surface, the gravitational force exerted on a body of mass m at a distance y above the surface is given by

$$F = -\frac{mgR^2}{(y + R)^2},$$

where the minus sign indicates that the force acts towards the planet's centre. The Cauchy problem for the equation of motion of an object of mass m projected upward from the surface reads

$$my'' = -\frac{mgR^2}{(y+R)^2} - c(y)(y')^2,$$
$$y(0) = R, \qquad y'(0) = v_0, \qquad (3.26)$$

where the last term in the equation represents the air resistance which, in this case, is taken to be proportional to the square of the velocity of the object and we allow the air resistance coefficient to change with height. The initial conditions tell us that the missile was shot from the surface with the initial velocity v_0.

Rather than solve the full Cauchy problem, we shall address the question of the existence of the *escape velocity*; that is, whether there exists an initial velocity which would allow the object to escape from the planet's gravitational field.

The equation in (3.26) is another example of the reducible equations discussed in Section 1.3.4. To simplify calculations, first we shall change the unknown function with the substitution to $z = y + R$ (so that z is the distance from the centre of the planet) and next introduce $F(z) = z'$ so that $z'' = F_z F$. Then the equation of motion takes the form

$$F_z F + C(z)F^2 = -\frac{gR^2}{z^2}, \qquad (3.27)$$

where $C(z) = c(z - R)/m$. Noting that $2F_z F = (F^2)_z$ and denoting $F^2 = H$, we reduce (3.27) to the inhomogeneous linear equation, discussed in Section 1.3.2,

$$H_z + 2C(z)H = -\frac{2gR^2}{z^2}. \qquad (3.28)$$

We shall consider three forms for C.

Case 1. $C(z) \equiv 0$ (an airless moon).
In this case (3.28) becomes $H_z = -2gR^2 z^{-2}$ which can be immediately integrated from R to z, giving

$$H(z) - H(R) = 2gR^2 \left(\frac{1}{z} - \frac{1}{R} \right).$$

Returning to the old variables $H(z) = F^2(z) = v^2(z)$, where v

is the velocity of a missile at a distance z from the centre of the moon, we can write

$$v^2(z) - v^2(R) = 2gR^2 \left(\frac{1}{z} - \frac{1}{R} \right).$$

The missile will escape from the moon if its speed remains positive for all times. In other words, if it stops at any finite z, then the gravity pull will bring it back to the moon. Since $v(z)$ is decreasing, its minimum value will be the limit at infinity so that, letting $z \to \infty$, we must have

$$v^2(R) \geq 2gR$$

and the escape velocity is

$$v(R) = \sqrt{2gR}. \tag{3.29}$$

Case 2. Constant air resistance.

If we are back on Earth, it is not reasonable to assume that there is no air resistance during the motion. Let us investigate the next simple case with constant c. Then we have

$$H_z + 2CH = -\frac{2gR^2}{z^2}, \tag{3.30}$$

where $C = c/m$. The integrating factor, see Section 1.3.2, is e^{2cz}, hence

$$\left(e^{2cz} H(z) \right)_z = -2gR^2 \frac{e^{2Cz}}{z^2}$$

and, upon integration,

$$v^2(z) = e^{-2Cz} \left(e^{2CR} v_0^2 - 2gR^2 \int_R^z e^{2Cs} s^{-2} ds \right). \tag{3.31}$$

Since $\lim_{s \to \infty} e^{2Cs} s^{-2} = \infty$, also $\lim_{z \to \infty} \int_R^z e^{2Cs} s^{-2} = \infty$. Since $\int_R^R e^{2Cs} s^{-2} ds = 0$ and because $e^{2CR} v^2(R)$ is independent of z, from the intermediate value theorem we see that, no matter what the value of v_0 is, for some $z_0 \in [R, \infty)$ the right hand side of (3.31) becomes 0 and thus $v^2(z_0) = 0$. Thus, there is no initial velocity v_0 for which the missile will escape the planet.

Case 3. Variable air resistance.

By passing from no air resistance at all $(c = 0)$ to a constant air resistance, we definitely overcompensated since the air becomes thinner with height and thus its resistance decreases. Let us consider one more case with $C(z) = k/z$, where k is a proportionality constant. Then we obtain

$$H_z + 2kz^{-1}H = -2gR^2z^{-2}.$$

The integrating factor is z^{2k}, hence we obtain

$$\left(z^{2k}H(z)\right)_z = -2gR^2z^{2k-2},$$

and, upon integration,

$$z^{2k}v^2(z) - R^{2k}v_0^2 = -2gR^2 \int_R^z s^{2k-2}ds.$$

Using the same argument as previously, we see that the escape velocity exists if and only if $\lim_{z \to \infty} \int_R^z s^{2k-2}ds < +\infty$ and, from the properties of improper integral, we infer that we must have $2k - 2 < -1$ or $k < \frac{1}{2}$. Of course, from physics, $k \geq 0$. Thus, the escape velocity in such a case is given by $v_0 = \sqrt{2gR/(1 - 2k)}$.

Clearly, apart from Case 1, which is perfectly applicable to airless objects, the other two models have been chosen for their mathematical tractability, so that we could present some examples of a more advanced mathematical reasoning.

We conclude by discussing some numerical results related to Case 1. Using (3.29), we find that the escape velocity from the Moon (no air, $R = 1737 \, \text{km}, g = 1.6 \, \text{m/s}^2$) is $v_0 = 2.3 \, \text{km/s}$. The escape velocity from the Earth must be larger than if the Earth was airless. Using $g = 9.81 \, \text{m/s}^2$ and $R \approx 6371 \text{km}$, we find $v_0 > 11.18 \, \text{km/s}$. For comparison, modern guns can propel bullets up to $1.7 \, \text{km/s}$ which is far too low to overcome the pull of gravity. Moreover, because of the atmosphere, it is not useful, and hardly possible, to give an object near the surface of the Earth the speed of $11.2 \, \text{km/s}$ as it would cause most objects to burn up due to atmospheric friction. This is the reason why currently deep space trips are divided into two stages. First, a multistage rocket, which provides a continual acceleration, is used to place a spacecraft in a low Earth orbit. Then, the spacecraft is accelerated to the escape velocity which at this altitude is much lower.

Exercise 3.5   A body of mass m moves from rest in a medium offering resistance proportional to the speed, that is, $D = cv$.

(a) Suitably modifying the derivation of (3.23), write the equation of motion of the body and the appropriate initial conditions.
(b) Find the velocity of the body $v(t)$ and compute its terminal velocity v_T.

3.5 Equations arising from geometrical modelling

3.5.1   Satellite dishes

In many applications, like radar or TV/radio transmission, it is important to find the shape of a surface that reflects parallel in-coming rays into a single point, called the focus. Conversely, con-structing a spotlight one needs a surface reflecting light rays com-ing from a point source to create a beam of parallel rays. To find an equation for a surface satisfying this requirement, we set the coordinate system so that the rays are parallel to the x-axis and the focus is at the origin. The surface we seek must have axial symmetry: that is, it must be a surface of revolution obtained by rotating some curve C about the x-axis. We have to find the equa-tion $y = y(x)$ of C. Using the notation of Fig. 3.5, we let $M(x, y)$ be an arbitrary point on the curve and denote by T the point at which the tangent to the curve at M intersects the x-axis. It follows that the triangle TOM is isosceles and

$$\tan \sphericalangle OTM = \tan \sphericalangle TMO = \frac{dy}{dx},$$

where the derivative is evaluated at P. By symmetry, we can as-sume that $y > 0$. Thus we can write $\tan \sphericalangle OTM = |MP|/|TP|$, but $|MP| = y$ and, since the triangle is isosceles, $|TP| = |OT| \pm |OP| = |OM| \pm |OP| = \sqrt{x^2 + y^2} + x$, irrespective of the sign of x. Thus, the differential equation of the curve C is

$$y' = \frac{y}{\sqrt{x^2 + y^2} + x}. \qquad (3.32)$$

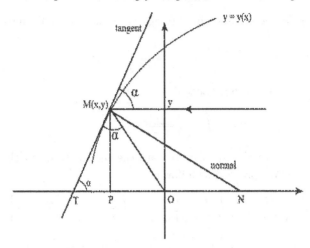

Figure 3.5 Geometry of a reflecting surface

Exercise 3.6 Modifying suitably the argument above, find the differential equation of a curve having the property that, given two points A and B not on the curve, any ray of light shone from A will be reflected by the curve to B.

It is interesting that the banqueting halls in many European medieval castles had the ceilings designed using the (3-dimensional) version of the above principle. The lord's table was placed at the point A while the table for the guests, whom the lord was interested to eavesdrop on, was placed at the point B, thus allowing him to hear what was being discussed there independently of the orientation of the speaker.

Observe that for $x > 0$ the right hand side of (3.32) can be written as

$$\frac{\frac{y}{x}}{\sqrt{1 + (\frac{y}{x})^2} + 1},$$

so that it is an equation of the homogeneous type, see Section 1.3.3. Thus it can be solved by the substitution $z = y/x > 0$.

Using $y' = z'x + z$ and simplifying, we transform (3.32) into

$$z'\left(\frac{1}{z} + \frac{1}{z\sqrt{z^2+1}}\right) = -\frac{1}{x}.$$

Integrating and using $z, x > 0$ we obtain

$$\ln z + \int \frac{dz}{z\sqrt{1+z^2}} = -\ln x + c'. \tag{3.33}$$

There are several ways to integrate the second term on the left hand side. We proceed as follows:

$$\int \frac{dz}{z\sqrt{1+z^2}} = \frac{1}{2}\int \frac{du}{u\sqrt{1+u}} = \int \frac{dv}{v^2-1} = \frac{1}{2}\ln\frac{v-1}{v+1}$$

$$= \frac{1}{2}\ln\frac{\sqrt{u+1}-1}{\sqrt{u+1}+1} = \frac{1}{2}\ln\frac{u}{(\sqrt{u+1}+1)^2}$$

$$= \ln z - \ln(1+\sqrt{z^2+1}),$$

where we used substitutions $u = z^2$, $v = \sqrt{u+1} \geq 1$ and also we omitted the constant of integration as it already appears in (3.33). Returning to (3.33), we obtain

$$\ln \frac{z^2}{1+\sqrt{z^2+1}} = -\ln x/c$$

for some constant $c > 0$. Thus

$$\frac{z^2}{1+\sqrt{z^2+1}} = \frac{c}{x},$$

and, returning to the original unknown function $y = zx$, after some algebra we get

$$y^2 - 2cx = c^2. \tag{3.34}$$

This is the equation of a parabola with vertex at $x = -c/2$ and with focus at the origin.

Exercise 3.7 Equation (3.34) was obtained under the assumption that $x > 0$ so, in fact, we do not have the full parabola at this moment. Show that (3.34) follows from (3.32) also under assumption that $x < 0$. How would you justify the validity of (3.34) for $x = 0$?

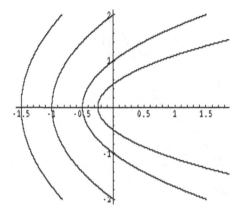

Figure 3.6 Different shapes of parabolic curves corresponding to various values of the constant c. In each case the focus is at the origin

3.5.2 The pursuit curve

What is the path of a dog chasing a rabbit or the trajectory of a self-guided missile trying to intercept an enemy plane? To answer this question we must first realize the principle used in controlling the chase. This principle is that at any instant the direction of motion (that is, the velocity vector) is directed towards the chased object.

To avoid technicalities, we assume that the target moves with a constant speed v along a straight line so that the pursuit takes place on a plane. We introduce the coordinate system in such a way that the target moves along the y-axis in the positive direction, starting from the origin at time $t = 0$, and the pursuer starts from a point at the negative half of the x-axis, see Fig. 3.7. We also assume that the pursuer moves with a constant speed u. Let $M = M(x(t), y(t))$ be a point on the curve C having the equation $y = y(x)$, corresponding to the time t of the pursuit at which $x = x(t)$. At this moment the position of the target is $(0, vt)$. From the principle of the pursuit we obtain

$$\frac{dy}{dx} = -\frac{vt - y}{x}, \tag{3.35}$$

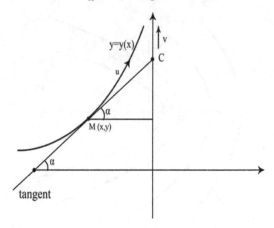

Figure 3.7 The pursuit curve

where we have taken into account that $x < 0$. In this equation we have too many variables and we shall eliminate t since we are looking for the equation of the trajectory in x, y variables. Solving (3.35) with respect to x, we obtain

$$x = -\frac{vt - y}{\frac{dy}{dx}}. \tag{3.36}$$

Using the assumption that v is constant, remembering that $x = x(t), y = y(x(t))$ and

$$\frac{dy}{dt} = \frac{dy}{dx}\frac{dx}{dt},$$

by differentiating (3.36) with respect to t, we get

$$\frac{dx}{dt} = \frac{\left(-v + \frac{dy}{dx}\frac{dx}{dt}\right)\frac{dy}{dx} + (vt - y)\frac{d^2y}{dx^2}\frac{dx}{dt}}{\left(\frac{dy}{dx}\right)^2}.$$

Multiplying out and simplifying yields

$$0 = -v\frac{dy}{dx} + (vt - y)\frac{d^2y}{dx^2}\frac{dx}{dt}$$

whereupon, using (3.36) and solving for $\frac{dx}{dt}$, we obtain

$$\frac{dx}{dt} = -\frac{v}{x\frac{d^2y}{dx^2}}. \tag{3.37}$$

On the other hand, we know that the speed of an object moving according to the parametric equation $(x(t), y(t))$ is given by

$$u = \sqrt{\left(\frac{dx}{dt}\right)^2 + \left(\frac{dy}{dt}\right)^2} = \sqrt{1 + \left(\frac{dy}{dx}\right)^2} \left|\frac{dx}{dt}\right|. \qquad (3.38)$$

From the formulation of the problem it follows that $\frac{dx}{dt} > 0$ (it would be unreasonable for the dog to start running away from the rabbit), hence we can drop the absolute value bars in (3.37). Thus, combining (3.37) and (3.38) we obtain the equation of the pursuit curve

$$x\frac{d^2y}{dx^2} = -\frac{v}{u}\sqrt{1 + \left(\frac{dy}{dx}\right)^2}. \qquad (3.39)$$

Though this is a second-order equation, it is reducible to one of first-order equations, see Section 1.3.4. To solve it, we introduce a new unknown by $z = dy/dx$ so that (3.39) becomes the first-order equation

$$x\frac{dz}{dx} = -k\sqrt{1 + z^2},$$

where we denoted $k = v/u$. This is a separable equation with a non-vanishing right hand side, so that we do not have stationary solutions. Separating variables and integrating, we obtain

$$\int \frac{dz}{\sqrt{1 + z^2}} = -k\ln(-C'x)$$

for some constant $C' > 0$, where we used the fact that $x < 0$ in the model. Integration (for example, substituting $z = 1/s$ and using the integral from the previous subsection) gives

$$\ln(z + \sqrt{z^2 + 1}) = \ln C(-x)^{-k},$$

with $C = (C')^{-k}$, hence

$$z + \sqrt{z^2 + 1} = C(-x)^{-k},$$

from where, after some algebra,

$$z = \frac{1}{2}\left(C(-x)^{-k} - \frac{1}{C}(-x)^k\right). \qquad (3.40)$$

Returning to the original unknown function y, where $dy/dx = z$, and integrating the above equation, we find

$$y(x) = \frac{1}{2}\left(\frac{1}{C(k+1)}(-x)^{k+1} - \frac{C}{(1-k)}(-x)^{-k+1}\right) + C_1.$$

Let us express the constants C_1 and C through the initial conditions. We assume that the pursuer started from the position $(x_0, 0)$, $x_0 < 0$, and that at the initial moment the target was at the origin $(0,0)$. Using the principle of the pursuit, we see that the initial direction was along the x-axis; that is, we obtain the initial conditions in the form

$$y(x_0) = 0, \qquad y'(x_0) = 0.$$

Since $dy/dx = z$, substituting $z = 0$ and $x = x_0$ in (3.40) yields

$$0 = \frac{dy}{dx}(x_0) = z(x_0) = C(-x_0)^{-k} - \frac{1}{C}(-x_0)^k$$

which gives $C = (-x_0)^k$, so that

$$y(x) = -\frac{x_0}{2}\left(\frac{1}{k+1}\left(\frac{x}{x_0}\right)^{k+1} - \frac{1}{1-k}\left(\frac{x}{x_0}\right)^{-k+1}\right) + C_1.$$

To find C_1, we let $x = x_0$ and $y(x_0) = 0$ above, getting

$$0 = -\frac{x_0}{2}\left(\frac{1}{k+1} + \frac{1}{k-1}\right) + C_1:$$

thus $C_1 = kx_0/(k^2 - 1)$. Finally,

$$y(x) = -\frac{x_0}{2}\left(\frac{1}{k+1}\left(\frac{x}{x_0}\right)^{k+1} - \frac{1}{1-k}\left(\frac{x}{x_0}\right)^{-k+1}\right) + \frac{kx_0}{k^2-1}.$$

This formula can be used to obtain two important pieces of information: the time and the point of interception. The interception occurs when $x = 0$. Thus

$$y(0) = \frac{kx_0}{k^2 - 1} = \frac{vux_0}{v^2 - u^2}.$$

The duration of the pursuit can be calculated by noting that the target moves with a constant speed v along the y-axis from the origin to the interception point $(0, y(0))$ so that

$$T = \frac{y(0)}{v} = \frac{ux_0}{v^2 - u^2}.$$

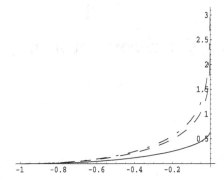

Figure 3.8 Pursuit curve for different values of k: $k = 0.5$ (solid line), $k = 0.9$ (dashed line), $k = 0.99$ (dot-dashed line)

Exercise 3.8 (a) Find the initial conditions if a dog spots a rabbit 20 m from the origin (in the positive direction along the y-axis) and hence find the solution of this problem.

(b) Let the dog spot the rabbit at the origin. The rabbit runs to the burrow which is 100 m away along the y-axis with a speed $v = 20$ km/h. What should be the smallest speed u of the dog if it is to catch the rabbit?

Exercise 3.9 When a tractor trailer turns into a cross street or driveway, its rear wheels follow a certain curve. Thus, to prevent a kerb at an intersection from being destroyed by trailers, it should be laid along this curve. Find the equation of this curve.

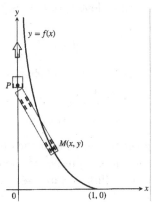

4

Qualitative theory for a single equation

In most cases it is impossible to find an explicit solution to a given differential, or difference, equation. However, the power of mathematics lies in the fact that one often can deduce properties of solutions and answer some relevant questions just by analyzing the right hand side of the equation.

4.1 Equilibria of first-order equations

One of the typical problems in the theory of differential and difference equations is to determine whether the system is stable; that is, whether if we allow it to run for a sufficiently long time (which, in the case of difference equations, means many iterations), it will eventually settle at some state, which should be an equilibrium.

In both difference and differential equations, by *equilibria* or *stationary solutions* we understand solutions which are constant with respect to the independent variable. Since, however, in the differential equation

$$x' = f(x) \tag{4.1}$$

the right hand side describes the rate of change of a given quantity, whereas in the difference equation

$$x_{n+1} = f(x_n) \tag{4.2}$$

the right hand side gives the amount of the quantity in the state $n + 1$ in relation to the amount present in the previous state, the theories are different and will be treated separately.

As we will see below, finding equilibria of equations is considerably easier than finding their general solution. Thus, knowing that the system will converge to a particular equilibrium allows us to regard this equilibrium as an approximation of solutions originating in its neighbourhood.

In the following subsections we will make these notions precise.

4.1.1 Stability of equilibria of autonomous differential equations

We recall that the word autonomous refers to the fact that f in (4.1) does not explicitly depend on time. To fix attention, we shall assume that f is an everywhere-defined function satisfying all assumptions of the Picard theorem on $\mathbb{R}$.

In many problems it is important to emphasize the dependence of the solution on the initial conditions. Thus we introduce the notion of the flow $x(t, t_0, x_0)$ of (4.1), which is the solution of the Cauchy problem

$$x'(t, t_0, x_0) = f(x(t, t_0, x_0)), \qquad x(t_0, t_0, x_0) = x_0.$$

Henceforth we take $t_0 = 0$ and write $x(t, 0, x_0) = x(t, x_0)$.

If (4.1) has a stationary solution $x(t) \equiv x^*$ that, by definition, is constant in time, then such a solution satisfies $x'(t) \equiv 0$ and consequently

$$f(x^*) = 0. \tag{4.3}$$

Conversely, if the equation $f(x) = 0$ has a solution, which we call an *equilibrium point* then, since f is independent of time, such a solution is a number, say x^*. If we now consider a function defined by $x(t) \equiv x^*$, then $x'(t) \equiv 0$. Consequently,

$$0 \equiv x'(t) \equiv (x^*)' = f(x^*)$$

and such a function is a stationary solution. Summarizing, equilibrium points are solutions of the algebraic equation (4.3) and, treated as constant functions, they are (the only) stationary, or equilibrium, solutions to (4.1). Therefore usually we will not differentiate between these terms.

Next we give a definition of stability of an equilibrium.

Definition 4.1   (i) The equilibrium x^* is *stable* if for given $\epsilon > 0$ there is $\delta > 0$ such that, for any x_0, $|x_0 - x^*| < \delta$ implies $|x(t, x_0) - x^*| < \epsilon$ for all $t > 0$. If x^* is not stable, then it is called *unstable*.

(ii) A point x^* is called *attracting* if there is $\eta > 0$ such that $|x_0 - x^*| < \eta$ implies $\lim_{t \to \infty} x(t, x_0) = x^*$. If $\eta = \infty$, then x^* is called a *global attractor* or a globally attracting equilibrium.

(iii) The equilibrium x^* is called *asymptotically stable* if it is both stable and attracting. If x^* is globally attracting, then it is said to be a *globally asymptotically stable equilibrium*.

Equilibrium points play another important role for differential equations – they are the only limit points of bounded solutions as $t \to \pm\infty$. To make this precise, we begin with the following lemma.

Lemma 4.2   *If x_0 is not an equilibrium point of* (4.1), *then $x(t, x_0)$ is never equal to an equilibrium point. In other words, $f(x(t, x_0)) \neq 0$ for any t for which the solution exists.*

Proof   An equilibrium point x^* generates a stationary solution, given by $x(t) \equiv x^*$. Thus, if $x(t_1, x_0) = x^*$ for some t_1, then (t_1, x_0) belongs to two different solutions, which contradicts the Picard theorem.   □

From the above lemma it follows that if f has several equilibrium points, then the stationary solutions corresponding to these points divide the (t, x) plane into horizontal strips having the property that any solution always remains confined to one of them. We shall formulate and prove a theorem that strengthens this observation.

Theorem 4.3   *Let $x(t, x_0)$ be a non-stationary solution of* (4.1) *with $x_0 \in \mathbb{R}$ and let $I_{\max} = (t_-, t_+)$ be its maximal interval of existence. Then $x(t, x_0)$ is either a strictly decreasing or a strictly increasing function of t. Moreover, $x(t, x_0)$ either diverges to $+\infty$ or to $-\infty$, or converges to an equilibrium point, as $t \to t_\pm$. In the latter case $t_\pm = \pm\infty$.*

Proof   Assume that for some $t_* \in I_{\max}$ the solution $x(t) :=$

$x(t, x_0)$ has a local maximum or minimum $x_* = x(t_*)$. Since $x(t)$ is differentiable, we must have $x'(t_*) = 0$ but then $f(x_*) = 0$ which makes x_* an equilibrium point of f. This means that a non-stationary solution $x(t, x_0)$ reaches an equilibrium in finite time, which contradicts Lemma 4.2. Thus, if $x(t, x_0)$ is not a stationary solution, then it cannot attain local maxima or minima and thus must be either strictly increasing or strictly decreasing.

Since the solution is monotonic, it either diverges to $\pm\infty$ (depending on whether it decreases or increases) or converges to finite limits as $t \to t_\pm$. Let us focus on the right end point t_+ of $I_{\max}$. If $x(t, x_0)$ converges as $t \to t_+$, then $t_+ = \infty$, by Theorem 1.6. Thus

$$\lim_{t\to\infty} x(t, x_0) = \bar{x}.$$

Without compromising generality, we further assume that $x(t, x_0)$ is an increasing function. If $\bar{x}$ is not an equilibrium point then, by continuity, we can use the intermediate value property to claim that the values of $x(t, x_0)$ must fill the interval $[x_0, \bar{x})$. This interval cannot contain any equilibrium point as the existence of such points would violate the Picard theorem. Thus, for any $x \le \bar{x}$, $f(x)$ is strictly positive and hence, separating variables and integrating, we obtain

$$t(x) - t(x_0) = \int_{x_0}^{x} \frac{ds}{f(s)}. \tag{4.4}$$

Passing with t to infinity (since $t(\bar{x}) = \infty$), we see that the left hand side becomes infinite and so

$$\int_{x_0}^{\bar{x}} \frac{ds}{f(s)} = \infty.$$

By assumption, the interval of integration is finite so that the only way the integral could become infinite is if $1/f(s) = \infty$; that is, $f(s) = 0$, for some $s \in [x_0, \bar{x}]$. The only such point can be $s = \bar{x}$, thus $\bar{x}$ is an equilibrium point. $\square$

Remark 4.4 We note that (4.4) is of independent interest as it gives a formula for the blow-up time of the solution $x(t, x_0)$. To wit, let the interval $[x_0, \infty)$ be free of equilibria and let $x(t, x_0)$ be increasing for $t > 0$. Then $\lim_{t \to t_+} x(t, x_0) = \infty$ so that, by (4.4),

$$t_+ - t(x_0) = \int\limits_{x_0}^{\infty} \frac{ds}{f(s)}$$

and, in particular, we see that if $1/f$ is integrable at $+\infty$ (precisely, if the improper integral above exists), then the maximal interval of existence is finite and we have the blow-up of the solution in finite time. On the other hand, if $1/f$ is not integrable, then $t_{\max} = +\infty$. We note that the latter occurs if $f(s)$ does not grow faster than s as $s \to \infty$. This occurs, e.g. if f is globally Lipschitz continuous; that is, if the Lipschitz constant in (1.29) can be chosen to be independent of the rectangle $\mathcal{R}$. This provides a sketch of the proof of the property mentioned after Example 1.10. If $f(s)$ behaves, say, as s^2 for large s, then the integral on the right hand side is finite and thus $t_{\max} < \infty$, see Exercise 1.9.

Exercise 4.5 Assume that $f : \mathbb{R} \to \mathbb{R}$ is Lipschitz continuous on each interval $[-a, a]$. Prove that if it is globally Lipschitz continuous, then there is M such that $|f(y)| \leq M|y|$ for any $y \in \mathbb{R}$. Show also that the converse statement does not hold.

Remark 4.6 It is important to emphasize that the assumption that f satisfies the assumptions of the Picard theorem everywhere is crucial. If there are non-Lipschitzian points, then the behaviour of solutions close to such points is not covered by Theorem 4.3, as we have seen in Example 1.10.

4.1.2 Application to the logistic equation

Consider the Cauchy problem for the logistic equation

$$y' = y(1 - y), \qquad y(0) = y_0. \tag{4.5}$$

We have solved this problem in Section 3.3.3. Let us now get as much information as possible about the solutions to this problem without actually solving it. First, we observe that the right hand side is given by $f(y) = y(1 - y)$, which is a polynomial, and therefore at each point of $\mathbb{R}^2$ the assumptions of Picard's theorem are satisfied; that is, only one solution of (4.5) passes through each point (t_0, y_0). However, f is not a globally Lipschitz function, so that this solution may be defined only on a finite time interval.

The equilibrium points are found solving $y(1 - y) = 0$, hence $y \equiv 0$ and $y \equiv 1$ are the only stationary solutions. Moreover, $f(y) < 0$ for $y < 0$ and $y > 1$ and $f(y) > 0$ for $0 < y < 1$. Hence, from Lemma 4.2, it follows that the solutions starting from $y_0 < 0$ will stay strictly negative, those starting from $0 < y_0 < 1$ will stay in this interval and those with $y_0 > 1$ will be larger than 1, for all times of their respective existence, as they cannot cross the equilibrium solutions. Then, from Theorem 4.3, we see that the solutions with negative initial condition are decreasing and therefore tend to $-\infty$ if time increases. In fact, they blow-up in finite time since, by integrating the equation, we obtain

$$t(y) = \int\limits_{y_0}^{y} \frac{d\eta}{\eta(1 - \eta)}$$

and we see, passing with y to $-\infty$ on the right hand side, that we obtain a finite time of the blow-up.

Next, solutions with $0 < y_0 < 1$ are bounded and thus, by Proposition 1.6, they are defined for all times. They are increasing and thus they must converge to the larger equilibrium; that is, $\lim_{t \to \infty} y(t, y_0) = 1$. Finally, if we start with $y_0 > 1$, then $y(t, y_0)$ decreases and thus is bounded from below, satisfying again $\lim_{t \to \infty} y(t, y_0) = 1$. The shape of the solution curves can be determined as in Section 3.3.3. By differentiating (4.5) with respect to time, we obtain

$$y'' = y'(1 - y) - yy' = y'(1 - 2y).$$

Since for each solution (apart from the stationary ones), y' has a fixed sign, we see that an inflection point can exist only for solutions starting at $y_0 \in (0, 1)$ and it occurs at $y = 1/2$, where the solution changes from being convex downward to being convex upward. In the two other cases, the second derivative is of constant sign, giving the solution convex upward for negative solutions and convex downward for solutions larger than 1.

We see that we got the same picture as when solving the equation but with much less work.

4.1.3 Crystal growth – a case study

In many applications, such as in photographic film production, it is important to be able to manufacture crystals of a given size. The process begins by adding a mixture of small crystals of various sizes to a certain solvent and keeping it mixed. Then we apply the so-called Ostwald ripening process, see (Friedman and Littman, 1994), which is based on the following observation: if one allows the process to continue for a long time, then either all crystal grains will dissolve into the solution, or all grains will become of the same size. Hence, one has to arrange the conditions, such as the concentration of the solution, for the second possibility to occur.

To start our modelling process, we will make some simplifying assumptions. First we assume that all crystals are of the same shape and differ only in size which can be described by a single real parameter, e.g., they may be boxes with edges (xa, xb, xc), where a, b, c are fixed reference dimensions and x is a real positive variable. If, instead of crystals, we apply the model to aerosols, we would think about balls with radius x. As our interest is in one-dimensional models, in what follows we assume that we deal with crystals of one size only and find conditions under which they will create crystals of the required size.

Consider a volume of fluid containing an amount of dissolved matter (solute) with (uniform) concentration $c(t)$ at time t. There is a saturation concentration c^*, which is the maximum solute per unit volume that the fluid can hold. If $c(t) > c^*$, then the excess solute precipitates out in solid (crystal) form. Actually, for the precipitation, $c(t)$ must be bigger than a certain quantity $c_x > c^*$ which depends on the size of the precipitating grains. The threshold constant c_x is given by the Gibbs–Thomson relation, $c_x = c^* e^{\Gamma/x}$, where Γ is a physical quantity that depends on the shape of the crystals, its material properties and the temperature (which here is assumed fixed). Hence, if $c(t) > c_x$, then the material will come out of the solution and will deposit onto the crystals, characterized by the size x, and if $c(t) < c_x$, then the material will dissolve back from the crystals. Using the Gibbs–Thompson relation, we define

$$L^*(t) = \frac{\Gamma}{\log \frac{c(t)}{c^*}}. \qquad (4.6)$$

Note that $L^*(t) < x$ if and only if $c(t) > c_x$, so that the crystals will grow if and only if $L^*(t) < x$. A semi-empirical law, see (Friedman and Littman, 1994), taking this observation into account yields the equation

$$x' = G(x, c(t)), \qquad (4.7)$$

where

$$G(x, c(t)) = \begin{cases} k_g \left(c(t) - c^* e^{\Gamma/x}\right)^g & \text{if} \quad x > L^*(t), \\ -k_d \left(c^* e^{\Gamma/x} - c(t)\right)^d & \text{if} \quad x < L^*(t), \end{cases}$$

and where k_g, k_d, g, d are positive constants with $1 \leq g, d \leq 2$. As expected, for $c(t) > c^*$ we have $x' > 0$; that is, the crystal grows. Conversely, for $c(t) < c^*$ we have $x' < 0$ and the crystal shrinks.

We assume that initially we have μ^* crystals of size $x(0) = x^*$. The crystals do not coalesce or split but may completely dissolve, in which case we have $x(t) = 0$ for some time t. The formula for $c(t)$ can be obtained as the sum of the initial concentration c_0 and the amount which was dissolved from the crystals initially present (in a unit volume):

$$c(t) = c_0 + \rho k_v \mu^* (x^*)^3 - \rho k_v \mu^* (x(t))^3, \qquad (4.8)$$

where k_v is a geometric parameter relating x^3 to the crystal volume ($k_v = abc$ in the case of a box and $k_v = 4\pi/3$ in the case of a sphere), and ρ is the mass density of the solid phase of the material. For further use, we introduce $\mu = \rho k_v \mu^*$ and $c_1 = c_0 + \mu(x^*)^3$. Note that c_1 is the total amount of the material per unit volume, in either crystal, or solution, form. Thus, with some abuse of notation, we can write

$$x' = G(x), \qquad x(0) = x^*, \qquad (4.9)$$

where

$$G(x) = \begin{cases} k_g \left(c_1 - \mu x^3 - c^* e^{\Gamma/x} \right)^g & \text{if} \quad c_1 - \mu x^3 > c^* e^{\Gamma/x}, \\ -k_d \left(c^* e^{\Gamma/x} - (c_1 - \mu x^3) \right)^d & \text{if} \quad c_1 - \mu x^3 < c^* e^{\Gamma/x}. \end{cases}$$

We observe that, since $g, d \geq 1$, G is continuously differentiable on each set $\{x > 0;\ c_1 - \mu x^3 - c^* e^{\Gamma/x} < 0\}$ and $\{x > 0;\ c_1 - \mu x^3 - c^* e^{\Gamma/x} > 0\}$. Thus, it is Lipschitz continuous for $x > 0$. The first question is to determine the equilibrium points. Denote $f(x) = c_1 - \mu x^3 - c^* e^{\Gamma/x}$ so that (4.9) can be written as

$$x' = \begin{cases} k_g(f(x))^g & \text{if} \quad f(x) > 0, \\ -k_d(-f(x))^d & \text{if} \quad f(x) < 0. \end{cases} \tag{4.10}$$

Lemma 4.7   *There exist at most two positive solutions of*

$$f(x) = c_1 - \mu x^3 - c^* e^{\Gamma/x} = 0. \tag{4.11}$$

Proof   We have $\lim_{x \to 0^+} f(x) = \lim_{x \to \infty} f(x) = -\infty$. Further,

$$f'(x) = -3\mu x^2 + \frac{\Gamma c^*}{x^2} e^{\Gamma/x}$$

and

$$f''(x) = -6\mu x - \frac{\Gamma c^*}{x^3} e^{\Gamma/x} - \frac{\Gamma^2 c^*}{x^4} e^{\Gamma/x},$$

so that $f''(x) < 0$ for all $x > 0$. Therefore, f' has at most one zero and thus, by Rolle's theorem, f can have at most two solutions. $\square$

In what follows we focus on the case when we have exactly two solutions, denoted $0 < \xi_1 < \xi_2$. Note that in practice this can be always achieved by taking the initial concentration c_0 large enough, so that $f(x_0) > 0$ for some chosen x_0. Then $\xi_1 < x_0 < \xi_2$.

Proposition 4.8

(i) *If $x^* > \xi_2$, then $x(t)$ decreases and $\lim_{t \to \infty} x(t) = \xi_2$.*

(ii) *If $\xi_1 < x^* < \xi_2$, then $x(t)$ increases and $\lim_{t \to \infty} x(t) = \xi_2$.*

(iii) *If $x^* < \xi_1$, then $x(t)$ decreases and there is finite time t_0 such that $x(t_0) = 0$.*

Proof Items (i) and (ii) follow directly from Theorem 4.3 (note that the solutions exist for all positive times as they are bounded). We have to reflect on (iii) as Theorem 4.3 is not directly applicable here (G does not satisfy assumptions of the Picard theorem on the whole real line; in fact, $x = 0$ clearly is not a point of Lipschitz continuity). However, the idea of the proof is still applicable. Indeed, clearly $x(t)$ decreases, that is, $x(t) \leq x^* < \xi_1$ for all $t \in [0, t_{\max})$ and, as f increases for $x < \xi_1$ we have $x'(t) = G(x(t)) \leq G(x^*) =: c < 0$. Thus, $x(t) \leq ct + x^*$ and $x(t_0) = 0$ for $t \leq -x^*/c$. □

The above result is a basis for the analysis of a more realistic case, where we have crystals of several sizes in the original solution, (Friedman and Littman, 1994). If, however, we are lucky to have crystals of only one size x^* and we want to produce crystals of a required size ξ_2 then, according to Proposition 4.8, we simply have to place these crystals in a solution with parameters c_1 and c^*, chosen so that ξ_2 is the larger root of (4.11) and x^* is larger than the other root ξ_1. Otherwise, if $x^* < \xi_1$, then the crystals will dissolve in the solution.

Exercise 4.9 Give an alternative proof of (iii) using (4.4), to estimate t_0 for which $x(t_0) = 0$.

4.2 Equilibrium points of difference equations

Consider the autonomous first-order difference equation

$$x_{n+1} = f(x_n), \qquad n \in \mathbb{N}_0, \tag{4.12}$$

with the initial condition x_0. It is clear that the solution to (4.12) is given by iterations

$$x_n = f(f(\cdots f(x_0))) = f^n(x_0) \tag{4.13}$$

and henceforth we will be using both notations.

A point x^* in the domain of f is said to be an *equilibrium point* of (4.2) if it is a fixed point of f; that is, if $f(x^*) = x^*$. In other words, the constant sequence $(x^*, x^*, \ldots)$ is a stationary solution of (4.2). As in the case of differential equations, here also we shall not differentiate between these concepts.

Example 4.10   Consider the logistic equation

$$x_{n+1} = 3x_n(1 - x_n). \tag{4.14}$$

The equation for the equilibrium points is $x = 3x(1 - x)$, which gives $x_0 = 0$ and $x_1 = 2/3$. Clearly, if $x_0 = 0$, then $x_n = 0$ for any $n \in \mathbb{N}$. Similarly, if $x_0 = 2/3$, then $x_1 = 3 \cdot (2/3) \cdot (1 - 2/3) = 2/3$ and, by iteration, $x_n = 2/3$ for any $n \in \mathbb{N}$.

Graphically, an equilibrium is the x-coordinate of any point where the graph of f intersects the diagonal $y = x$. This is the basis of the cobweb method of finding and analysing equilibria, described in the next subsection.

Definition 4.11   1. The equilibrium x^* is *stable* if for given $\epsilon > 0$ there is $\delta > 0$ such that for any x and for any $n > 0$, $|x - x^*| < \delta$ implies $|f^n(x) - x^*| < \epsilon$ for all $n > 0$. If x^* is not stable, then it is called *unstable* (that is, x^* is unstable if there is $\epsilon > 0$ such that for any $\delta > 0$ there are x and n such that $|x - x^*| < \delta$ and $|f^n(x) - x^*| \geq \epsilon$.)

2. A point x^* is called *attracting* if there is $\eta > 0$ such that $|x_0 - x^*| < \eta$ implies $\lim_{n \to \infty} f^n(x_0) = x^*$. If $\eta = \infty$, then x^* is called a *global attractor* or *globally attracting*.

3. The point x^* is called an *asymptotically stable equilibrium* if it is stable and attracting. If $\eta = \infty$, then x^* is said to be a *globally asymptotically stable equilibrium*.

4.2.1 The cobweb diagrams

We describe an important graphical method, the so-called cobweb diagrams, for analysing the stability of equilibrium (and periodic) points of (4.2). Since $x_{n+1} = f(x_n)$, we may draw a graph of f in the (x_n, x_{n+1}) system of coordinates. Then, given x_0, we pinpoint the value x_1 by drawing a vertical line through x_0 so that it also intersects the graph of f at (x_0, x_1). Next, we draw a horizontal line from (x_0, x_1) to meet the diagonal line $y = x$ at the point (x_1, x_1). A vertical line drawn from the point (x_1, x_1) will meet the graph of f at the point (x_1, x_2). In this way we may find any x_n. This is illustrated in Fig. 4.1, where we present several steps of drawing the cobweb diagram for the logistic equation (4.14)

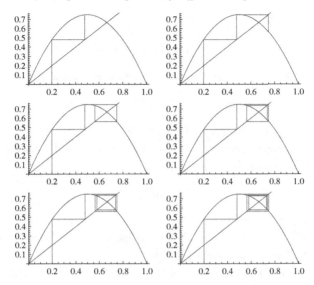

Figure 4.1 Cobweb diagram of a logistic difference equation

with $x_0 = 0.2$. On the basis of the diagram we can conjecture that $x_1 = 2/3$ is an asymptotically stable equilibrium as the solution converges to it as n becomes large. However, to be sure, we need to develop analytical tools for analysing stability.

4.2.2 Analytic criterion for stability

Theorem 4.12 *Let x^* be an isolated equilibrium point of the difference equation*

$$x_{n+1} = f(x_n), \qquad (4.15)$$

where f is continuously differentiable in some neighbourhood of x^*. Then,

(i) *if $|f'(x^*)| < 1$, then x^* is asymptotically stable;*
(ii) *if $|f'(x^*)| > 1$, then x^* is unstable.*

Proof Suppose $|f'(x^*)| < M < 1$. Then $|f'(x)| \leq M < 1$ over some interval $J = (x^* - \gamma, x^* + \gamma)$ by the property of local preservation of sign for continuous functions, (Courant and John, 1999).

Let $x_0 \in J$. We have

$$|x_1 - x^*| = |f(x_0) - f(x^*)|$$

and, by the mean value theorem, for some $\xi \in [x_0, x^*]$,

$$|f(x_0) - f(x^*)| = |f'(\xi)||x_0 - x^*|.$$

Hence, $|x_1 - x^*| = |f(x_0) - f(x^*)| \le M|x_0 - x^*|$. Since $M < 1$, the inequality shows that x_1 is closer to x^* than x_0 and consequently $x_1 \in J$. By induction,

$$|x_n - x^*| \le M^n |x_0 - x^*|.$$

For given ϵ, define $\delta = \epsilon$. Then $|x_n - x^*| < \epsilon$ for $n > 0$ provided $|x_0 - x^*| < \delta$ (since $M < 1$). Furthermore $x_n \to x^*$ and $n \to \infty$ so that x^* is asymptotically stable.

To prove the second part of the theorem, we observe that, as in the first part, there is $\epsilon > 0$ such that on $J = (x^* - \epsilon, x^* + \epsilon)$ we have $|f'(x)| \ge M > 1$. Take an arbitrary $\delta > 0$ smaller than ϵ and choose x satisfying $|x - x^*| < \delta$. Again using the mean value theorem, we get $|f(x) - x^*| = |f'(\xi)||x - x^*|$ for some ξ between x^* and x so that $|f(x) - x^*| \ge M|x - x^*|$. If $f(x)$ is outside J, then we are done. If not, we can repeat the argument getting $|f^2(x) - x^*| \ge M^2 |x - x^*|$; that is, $f^2(x)$ is further away from x^* than $f(x)$. If $f^2(x)$ is still in J, we continue the procedure till $|f^n(x) - x^*| \ge M^n |x - x^*| > \epsilon$ for some n. $\qquad \square$

Equilibrium x^* with $|f'(x^*)| \ne 1$ is called *hyperbolic*.

What happens if the equilibrium point x^* is not hyperbolic? To simplify the considerations, we assume that f is at least three times continuously differentiable in a neighbourhood of x^*. First, let us reflect on the geometry of the situation. In this discussion we assume that $f'(x^*) > 0$. The equilibrium x^* is stable if the graph of $y = f(x)$ less steep than the graph of $y = x$; that is, if the graph of f crosses the line $y = x$ from above to below as x increases. This ensures that the cobweb iterations from the left are increasing, and from the right are decreasing, while converging to x^*. In contrast, x^* is unstable if the graph of f crosses $y = x$ from below – then the cobweb iterations will move away from x^*. If $f'(x^*) = 1$, then the graph of f is tangent to the line $y = x$

at $x = x^*$, but the stability properties follow from the geometry. If $f''(x^*) \neq 0$, then f is convex (resp. concave) close to x^* and the graph of f will be (locally) entirely above (resp. entirely below) the line $y = x$. Therefore the picture is the same as in the unstable case either to the left, or to the right, of x^*. Hence, x^* is unstable in this case (remember that for instability it is sufficient to display, for any neighbourhood of x^*, only one diverging sequence of iterations emanating from this neighbourhood). On the other hand, if $f''(x^*) = 0$, then x^* is an inflection point and the graph of f crosses the line $y = 0$. This case is essentially the same as when $|f'(x^*)| \neq 1$: the equilibrium is stable if the graph of f crosses $y = x$ from above and unstable if it does it from below. A quick reflection ascertains that the former occurs when $f'''(x^*) < 0$, while the latter if $f'''(x^*)$. Summarising, we have:

Theorem 4.13 *Let x^* be an isolated equilibrium with $f'(x^*) = 1$ and let f be at least three times continuously differentiable in a neighbourhood of x^*. Then:*

(i) *if $f''(x^*) \neq 0$, then x^* is unstable;*

(ii) *if $f''(x^*) = 0$ and $f'''(x^*) > 0$, then x^* is unstable;*

(iii) *if $f''(x^*) = 0$ and $f'''(x^*) < 0$, then x^* is asymptotically stable.*

The case of $f'(x^*) = -1$ is more difficult. First we note that if $g(x) = -x + 2x^*$ – that is, if g is a linear function giving an equilibrium at $x = x^*$ with $f'(x^*) = -1$ – then the iterations starting from $x_0 \neq x^*$ produce a solution taking on only two values oscillating around x^*. Thus, if $-1 < f'(x^*) < 0$, then f passes from below the line $y = -x + 2x^*$ to above as x increases. Hence, the stability follows from the fact that subsequent iterations oscillate around x^* getting closer to x^* with each iteration. If, on the other hand, $f'(x^*) < -1$, then the oscillating iterations move away from x^*. If $f'(x^*) = -1$, then the graph of f crosses the line $y = x$ at a right angle. Hence, the stability depends on fine detail of the shape of f close to x^*. Unfortunately, using an argument similar to the case with $f'(x^*) = 1$ and, considering the relation of the graph of f with the graph of $y = -x + 2x^*$, only produces a partial result: x^* will be stable if $f''(x^*) = 0$ and $f'''(x^*) > 0$

(because then the graph of f will have the same shape as in the stable case, crossing the line $y = -x + 2x^*$ from below). However, the stability of x^* can be achieved in a more general situation. First, we note that x^* is also an equilibrium of $g(x) := f(f(x))$ and it is a stable equilibrium of f if and only if it is stable for g. This statement follows from the continuity of f: if x^* is stable for g, then $|g^n(x_0) - x^*| = |f^{2n}(x_0) - x^*|$ is small for x_0 sufficiently close to x^*. But then $|f^{2n+1}(x_0) - x^*| = |f(f^{2n})(x_0) - f(x^*)|$ is also small by continuity of f. The reverse is obvious. Since $g'(x) = f'(f(x))f'(x)$ with $g'(x^*) = 1$, we can apply Theorem 4.13 to the function g. The second derivative of g is given by

$$g''(x) = f''(f(x))[f'(x)]^2 + f'(f(x))f''(x)$$

and, since $f(x^*) = x^*$ and $f'(x^*) = -1$, we have $g''(x^*) = 0$. Using again the chain rule, we find

$$g'''(x^*) = -2f'''(x^*) - 3[f''(x^*)]^2.$$

Hence, we can write

Theorem 4.14   *Suppose that at an equilibrium point x^* we have $f'(x^*) = -1$. Define $S(x^*) = -f'''(x^*) - 3(f''(x^*))^2/2$. Then x^* is asymptotically stable if $S(x^*) < 0$ and unstable if $S(x^*) > 0$.*

Example 4.15   Consider the equation

$$x_{n+1} = x_n^2 + 3x_n.$$

Solving $f(x) = x^2 + 3x = x$, we find that $x = 0$ and $x = -2$ are the equilibrium points. Since $f'(0) = 3 > 1$, we conclude that the equilibrium at $x = 0$ is unstable. Next, $f'(-2) = -1$. We calculate $f''(-2) = 2$ and $f'''(-2) = 0$ so that $S(-2) = -12 < 0$. Hence, $x = -2$ is an asymptotically stable equilibrium.

Exercise 4.16   Consider the equation $x_{n+1} = Tx_n$, where

$$T(x) = \begin{cases} 2x & \text{for } 0 \le x \le 1/2, \\ 2(1-x) & \text{for } 1/2 < x \le 1, \end{cases}$$

is the so-called tent map. Find the equilibria and determine their stability.

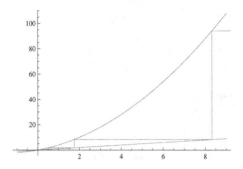

Figure 4.2 Unstable character of the equilibrium $x = 0$ in Example 4.15. Initial point $x_0 = 0.5$

Remark 4.17 We can fine tune the notion of stability by noting that if $f'(x^*) < 0$, then the solution behaves in an oscillatory way around x^* and if $f'(x^*) > 0$, then it is monotonic. Indeed, consider (in a neighbourhood of x^* where $f'(x) < 0$)

$$f(x) - f(x^*) = f(x) - x^* = f'(\xi)(x - x^*), \quad \xi \in (x^*, x).$$

Since $f' < 0$, $f(x) > x^*$ if $x < x^*$ and $f(x) < x^*$ if $x > x^*$, hence each iteration moves the point to the other side of x^*. If $|f'| < 1$ over this interval, then $f^n(x)$ converges to x^* in an oscillatory way, while if $|f'| > 1$, the iterations will move away from the interval, also in an oscillatory way.

Based on this observation, we may say that the equilibrium is *oscillatory unstable* or *stable* if $f'(x^*) < -1$ or $-1 < f'(x^*) < 0$, respectively, and *monotonically stable* or *unstable* depending on whether $0 < f'(x^*) < 1$ or $f'(x^*) > 1$, respectively.

Exercise 4.18 Consider the Allee model (2.22).

(i) Investigate stability of the equilibria 0, L and K in this model.
(ii) Show that if $N_0 > 0$, then $N_k > 0$ for all $k \in \mathbb{N}$, provided $0 < r < \min\{1/KL, 1/K(K - L)\}$.

4.2.3 Periodic points and cycles

Theorem 4.3 tells us that a solution to a scalar autonomous differential equation must be monotonic. We have already seen that

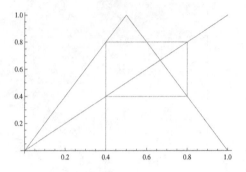

Figure 4.3 2-cycle for the tent map

solutions to scalar autonomous difference equations can be oscillatory. In fact, such equations may admit periodic solutions which cannot occur in the continuous case.

Definition 4.19   Let b be a point in the domain of f.

(i) We call b a *periodic point* of f if $f^k(b) = b$ for some $k \in \mathbb{N}$. The periodic orbit of b, $O(b) = \{b, f(b), f^2(b), \ldots, f^{k-1}(b)\}$ is called a *k-cycle*. (ii) We say b is *eventually k-periodic* if, for some integer m, $f^m(b)$ is a k-periodic point.

Example 4.20   Consider $x_{n+1} = T^2 x_n$, where T is the tent map introduced in Exercise 4.16. Then

$$T^2(x) = \begin{cases} 4x & \text{for} \quad 0 \le x \le 1/4, \\ 2(1-2x) & \text{for} \quad 1/4 < x \le 1/2, \\ 2x - 1 & \text{for} \quad 1/2 < x \le 3/4, \\ 4(1-x) & \text{for} \quad 3/4 < x \le 1. \end{cases}$$

There are four equilibrium points, $0, 2/5, 2/3$ and $4/5$, two of which are equilibria of T. Hence $\{2/5, 4/5\}$ is the only 2-cycle of T.

Definition 4.21   Let b be a k-periodic point of f. Then b is said to be:

(i) *stable*, if it is a stable fixed point of f^k;
(ii) *asymptotically stable*, if it is an asymptotically stable fixed point of f^k;
(iii) *unstable*, if it is an unstable fixed point of f^k.

This definition, together with Theorem 4.12, yields the following classification of the stability of k-cycles.

Theorem 4.22 *Let $O(b) = \{x_0 = b, x_1 = f(b), \ldots, x_{k-1} = f^{k-1}(b)\}$ be a k-cycle of a continuously differentiable function f. Then:*

(i) *the k-cycle $O(b)$ is asymptotically stable if*

$$|f'(x_0)f'(x_1)\cdots f'(x_{k-1})| < 1;$$

(ii) *the k-cycle $O(b)$ is unstable if*

$$|f'(x_0)f'(x_1)\cdots f'(x_{k-1})| > 1.$$

Proof Follows from Theorem 4.12 by the Chain Rule applied to f^k. □

Exercise 4.23 Determine the stability of the 2-cycle of the tent map.

4.2.4 The logistic equation and bifurcations

Consider the logistic equation

$$x_{n+1} = F_\mu(x_n) := \mu x_n(1 - x_n), \quad x \in [0,1],\ \mu > 0. \qquad (4.16)$$

Our aim is to investigate the properties of equilibria of (4.16) with respect to the parameter μ. The values of μ for which there is a qualitative change of the properties of the equilibria are called *bifurcation points*.

To find the equilibrium points, we solve $F_\mu(x^*) = x^*$ which gives $x^* = 0, (\mu - 1)/\mu$.

We investigate the stability of each point separately.

(a) For $x^* = 0$, we have $F_\mu'(0) = \mu$ and thus $x^* = 0$ is asymptotically stable for $0 < \mu < 1$ and unstable for $\mu > 1$. To investigate the stability for $\mu = 1$, we find $F_\mu''(0) = -2 \neq 0$ and thus $x^* = 0$ is unstable in this case. However, the instability comes from the negative values of x, which we discarded from the domain. If we restrict our attention to the domain $[0,1]$, then $x^* = 0$ is stable. Such points are called *semi-stable*.

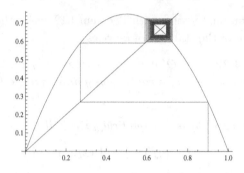

Figure 4.4 Asymptotically stable equilibrium $x = 2/3$ for $\mu = 3$.

(b) The equilibrium point $x^* = (\mu - 1)/\mu$ belongs to the domain $[0,1]$ only if $\mu > 1$. Here, $F'((\mu - 1)/\mu) = 2 - \mu$ and $F''((\mu - 1)/\mu) = -2\mu$. Thus, using Theorems 4.12 and 4.13, we obtain that x^* is asymptotically stable if $1 < \mu \le 3$ and it is unstable if $\mu > 3$.

Further, by Remark 4.17, we observe that for $1 < \mu < 2$, the population approaches the carrying capacity monotonically from below. However, for $2 < \mu \le 3$ the population can go over the carrying capacity but it will eventually stabilize around it.

Exercise 4.24 Determine whether the solution for $\mu = 2$ is monotonic.

What happens for $\mu = 3$? Consider 2-cycles. We have $F_\mu^2(x) = \mu^2 x(1 - x)(1 - \mu x(1 - x))$ so that we are looking for solutions to $\mu^2 x(1 - x)(1 - \mu x(1 - x)) = x$, which can be rewritten as

$$x(\mu^3 x^3 - 2\mu^3 x^2 + \mu^2(1 + \mu)x + (1 - \mu^2)) = 0.$$

To simplify, we observe that any equilibrium is also a 2-cycle (and any k-cycle for that matter). Thus, we can divide this equation by x and $x - (\mu - 1)/\mu$, getting

$$\mu^2 x^2 - \mu(\mu + 1)x + \mu + 1 = 0.$$

Solving this quadratic equation, we obtain a 2-cycle

$$x_\pm = \frac{(1 + \mu) \pm \sqrt{(\mu - 3)(\mu + 1)}}{2\mu}. \tag{4.17}$$

Clearly, these points determine a 2-cycle provided $\mu > 3$ (in fact, for $\mu = 3$, these two points collapse into the equilibrium point $x^* = 2/3$. Thus, we see that when the parameter μ passes through $\mu = 3$, the stable equilibrium becomes unstable and bifurcates into two 2-cycles.

The stability of 2-cycles can be determined by Theorem 4.22. We have $F'(x) = \mu(1 - 2x)$ so the 2-cycle is stable, provided

$$-1 < \mu^2(1 - 2x_+)(1 - 2x_-)) < 1.$$

Using Viète's formulae we find that the above is satisfied provided $-1 < -\mu^2 + 2\mu + 4 < 1$ and, upon solving this inequality, we get $\mu < -1$ or $\mu > 3$ and $1 - \sqrt{6} < \mu < 1 + \sqrt{6}$ which yields $3 < \mu < 1 + \sqrt{6}$.

In a similar fashion, we find that for $\mu_1 = 1 + \sqrt{6}$, the 2-cycle is still attracting but becomes unstable for $\mu > \mu_1$.

To find 4-cycles, we solve $F_\mu^4(x) = 0$. However, in this case the algebra becomes analytically intractable and we should resort to numerical approximations. It turns out that there is a 4-cycle, when $\mu > 1 + \sqrt{6}$, which is attracting for $1 + \sqrt{6} < \mu < 3.544090\ldots =: \mu_2$. When $\mu = \mu_2$, the 2^2-cycle bifurcates into a 2^3-cycle which is stable for $\mu_2 \leq \mu \leq \mu_3 := 3.564407\ldots$. Continuing, we obtain a sequence of numbers (μ_n), such that the 2^n-cycle bifurcates into a 2^{n+1}-cycle passing through μ_n. In this particular case, $\lim_{n\to\infty} \mu_n = \mu_\infty = 3.57\ldots$. A remarkable observation, made by Feigenbaum, is that for any sufficiently smooth family F_μ of mappings of an interval into itself, the number

$$\delta = \lim_{n\to\infty} \frac{\mu_n - \mu_{n-1}}{\mu_{n+1} - \mu_n} = 4.6692016\ldots,$$

in general does not depend on the family of maps, provided they have single maximum. The interested reader should consult e.g. (Elaydi, 2005) for further information on dynamics of the logistic map.

Feigenbaum's result expresses the fact that the picture obtained for the logistic equation is to a large extent universal. What happens for μ_∞? Here we find a densely interwoven region with both periodic and very irregular orbits. In particular, a 3-cycle appears and, by a celebrated theorem of Šarkovsky, see e.g. (Glendinning,

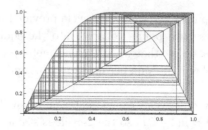

Figure 4.5 Chaotic orbit for $x = 0.9$ and $\mu = 4$.

1994), this implies the existence of orbits of any period. In fact, what we observe is the emergence of the so-called chaotic dynamics. The discussion of this topic is beyond the scope of this book and the reader is referred to e.g. (Elaydi, 2005) (discrete dynamics) and (Glendinning, 1994; Strogatz, 1994) (continuous dynamics) for a more detailed account of it.

Exercise 4.25   Show that 3-cycles appear if $\mu = 4$.

4.2.5 Stability in the Beverton–Holt equation

We conclude with a brief description of the stability of equilibrium points for the Beverton–Holt equation which was discussed in Subsections 2.2.4 and 2.4.1. Let us recall this equation

$$x_{n+1} = f(x_n, R_0, b) = \frac{R_0 x_n}{(1 + x_n)^b}.$$

Writing $x^*(1 + x^*)^b = R_0 x^*$, we find a steady state $x^* = 0$ and we observe that if $R_0 \leq 1$, then this is the only steady state (at least for positive values of x). If $R_0 > 1$, then there is another steady state given by $x^* = R_0^{1/b} - 1$. Evaluating the derivative at x^*, we have

$$f'(x^*, R_0, b) = \frac{R_0}{(1 + x^*)^b} - \frac{R_0 b x^*}{(1 + x^*)^{b+1}} = 1 - b + \frac{b}{R_0^{1/b}}.$$

Clearly, with $R_0 > 1$, we always have $f' < 1$. Hence, for monotone stability we must have $1 - b + b R_0^{-1/b} > 0$, whereas oscillatory stability requires $-1 < 1 - b + b R_0^{-1/b} < 0$. Solving these inequal-

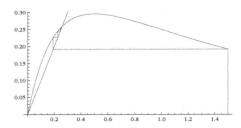

Figure 4.6 Monotonic stability of the equilibrium for the Beverton–Holt model with $b = 3$ and $R_0 = 2$; see (4.18).

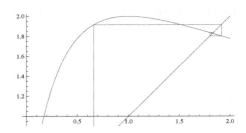

Figure 4.7 Oscillatory stability of the equilibrium for the Beverton–Holt model with $b = 2$ and $R_0 = 8$; see (4.18).

ities, we obtain that the borderlines between different types of behaviour are given by

$$R_0 = \left(\frac{b}{b-1}\right)^b \text{ and } R_0 = \left(\frac{b}{b-2}\right)^b. \qquad (4.18)$$

Let us consider the existence of 2-cycles. The second iteration of the map $H(x) = R_0 x/(1+x)^b$ is given by

$$H(H(x)) = \frac{R_0^2 x (1+x)^{b^2-b}}{((1+x)^b + R_0 x)^b},$$

so that 2-cycles can be obtained by solving $H(H(x)) = x$. This can be rewritten as

$$x R_0^2 (1+x)^{b^2-b} = x((1+x)^b + R_0 x)^b,$$

or, discarding $x = 0$ and taking the bth root,

$$(1+x)^{b-1} R_0^{2/b} = (1+x)^b + R_0 x.$$

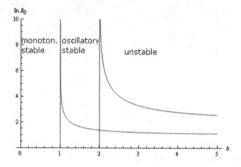

Figure 4.8 Regions of stability of the Beverton–Holt model described by (4.18)

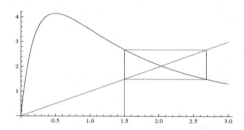

Figure 4.9 2-cycles for the Beverton-Holt model with $b = 3$ and $R_0 = 28$; see Eqn (4.18).

Introducing the change of variables $z = 1 + x$, we see that we have to investigate the existence of positive roots of

$$f(z) = z^b - z^{b-1}R_0^{2/b} + R_0 z - R_0.$$

Clearly, we have $f(R_0^{1/b}) = 0$, since any equilibrium of H is also an equilibrium of H^2. First, let us consider $1 < b \leq 2$ (the case $b = 1$ yields an explicit solution obtained in Section 2.4).

We have $f'(z) = bz^{b-1} - (b-1)z^{b-2}R_0^{2/b} + R_0$ and $f''(z) = (b-1)z^{b-3}(bz + (2-b)R_0^{2/b})$, hence we see that $f'' > 0$ for all $z > 0$. Furthermore, $f(0) = -R_0 < 0$. Hence, the region Ω, bounded from the left by the line $z = 0$ and lying above the graph of f for $z > 0$, is convex. Thus, the z-axis, being perpendicular to the line $z = 0$, cuts the boundary of Ω in exactly two points, one being $(0,0)$ and the other $(R_0^{1/b}, 0)$. Hence, there are no additional equilibria of H^2 for $1 < b < 2$ and therefore H does not have 2-cycles for $b \leq 2$.

Exercise 4.26 Prove the above statement directly if $b = 2$.

Consider next $b > 2$. Then f'' has exactly one positive root $z_i = (b-2)R_0^{2/b}/b$. The fact that the equilibrium $x^* = R_0^{1/b} - 1$ loses stability at $R_{0,\text{crit}} = (b/(b-2))^b$ suggests that a 2-cycle can appear, when R_0 increases passing through this point.

The analysis of this case rests on the following observation.

Lemma 4.27 *Let g be a differentiable function on an interval $[\alpha, \beta]$ such that $g(\alpha) > 0$, $g(\beta) \geq 0$ and the only zero of g' is at $x = \alpha$ or at $x = \beta$. Then $g(x) > 0$ for all $x \in [\alpha, \beta)$.*

Proof Assume there is $\alpha < x_0 < \beta$ where $g(x_0) \leq 0$. It cannot be the local minimum of g as then $g'(x_0) = 0$, contrary to the assumption. Thus, for some $x_1 > x_0$ we have $g(x_1) < 0$ and therefore there is $x_2 \in (x_1, \beta]$ with $g(x_2) = 0$ and, by Rolle's theorem, there is $x_3 < \beta$ for which $g'(x_3) = 0$, again contradicting the assumption. $\qquad\square$

Let us discuss the stable region $R_0 \leq b^b(b-2)^{-b}$. Then $z_i \leq (b-2)/2 = R_0^{1/b}$, that is, z_i is below the original equilibrium $R_0^{1/b}$; it is easy to see that, when R_0 passes through the value $R_{0,\text{crit}}$, z_i passes through $R_0^{1/b}$ and moves on to take on values larger than the original equilibrium. Let us evaluate the first derivative at z_i

$$f'\left(\frac{b-2}{b}R_0^{2/b}\right) = R_0\left(1 - \left(\frac{b-2}{b}\right)^{b-2}R_0^{(b-2)/b}\right).$$

Thus we see that $f'\left(\frac{b-2}{b}R_0^{2/b}\right) > 0$ provided $z_i < R_0^{1/b}$ and becomes negative as z_i moves through $R_0^{1/b}$. Furthermore, we see that $f'(R_0^{1/b}) = R_0(bR_0^{-1/b} - (b-2))$ and $f'(R_0^{1/b}) > 0$ provided $R_0 < (b(b-2))^b$, that is, for $z_i < R_0^{1/b}$.

Now, consider the case with $R_0 < (b(b-2))^b$. Since then $f'(0) = R_0 > 0$, $f'\left(\frac{b-2}{b}R_0^{2/b}\right) > 0$, $f'(R_0^{1/b}) > 0$, $0 < \frac{b-2}{b}R_0^{2/b} < R_0^{1/b}$ and $\frac{b-2}{b}R_0^{2/b}$ is the only zero of f'' on $[0, R_0^{1/b}]$, we can apply the lemma above (to $g = f'$ on the intervals $[0, z_i]$ and $[z_i, R_0^{1/b}]$). Hence, f' is positive on the interval $[0, R_0^{1/b}]$ and $R_0^{1/b}$ is the only zero of f in this interval. Consider now the interval $[R_0^{1/b}, \infty)$. Since $f'(R_0^{1/b}) > 0$ and $f(z)$ tends to $+\infty$ for $z \to \infty$, for f to have

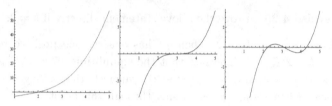

Figure 4.10 Function f for $b = 3$ and, from left to right, $R_0 = 8, 27, 30$. Notice the emergence of 2-cycles represented here by new zeroes of f besides $z = \sqrt[3]{R_0}$.

zeroes in this interval, it would have to have a local maximum, then take on 0, and then to have a local minimum before crossing the axis again to move to infinity. This would give another zero of f'' between the local extrema, but then this zero would be greater than $R_0^{1/b}$, which is impossible.

For $R_0 = (b(b-2))^b$, the points z_i and $R_0^{1/b}$ coalesce and, analogously, we see that there is only one zero of f.

Let us consider the case with $R_0 > (b(b-2))^b$. Then $f'(R_0^{1/b}) < 0$ and hence f takes on negative values for $z > R_0^{1/b}$. Since, however, $f(z)$ tends to $+\infty$ for $z \to \infty$, there must be $z^* > R_0^{1/b}$, for which $f(z^*) = 0$. Also, by $f'(R_0^{1/b}) < 0$, $f(z) > 0$ in a left neighbourhood of $R_0^{1/b}$ and hence there must be another zero $1 < z^\# < R_0^{1/b}$, by $f(1) = 1 - R_0^{1/b} < 0$. Since $R_0^{1/b} - 1$ and 0 were the only equilibria of H, $x^* = z^* - 1$ and $x^\# = z^\# - 1 > 0$ must give a 2-cycle.

Figure 4.10 shows, for $b = 3$, how the point z_i moves with R_0 through the equilibrium point $z = 3$ to produce new zeroes of f, giving rise to 2-cycles. With much more, mainly computer aided, work we can establish that, as with the logistic equation, we obtain period doubling and transition to chaos.

The Beverton–Holt models is mostly used to describe insect or fish populations. Experiments are in quite good agreement with the model, see (Britton, 2003). Most models fall into the stable region. On the other hand, it is obvious that a high reproductive ratio R_0 and highly over-compensating density dependence (large b) are capable of provoking periodic or chaotic fluctuations in the population density. This can be demonstrated mathematically (before the advent of the mathematical theory of chaos it was assumed that these irregularities were of stochastic nature) and it is observed in the fluctuations of the populations of the Colorado beetle.

The question of whether chaotic behaviour does exist in ecology is still an area of active debate. Observational time series are always finite and inherently noisy, thus it can be argued that always a regular model can be found to fit the data. However, in several laboratory host–parasitoid systems, good fits were obtained between the data and chaotic mathematical models and therefore it is reasonable to treat these systems as chaotic.

5

From discrete to continuous models and back

Unless a given phenomenon occurs at well-defined and evenly-spaced time intervals, it is up to us whether we describe it using difference or differential equations. Each choice has both advantages and disadvantages which, however, are closely intertwined. Indeed, as we have seen, continuous models are obtained using the same principles as the corresponding discrete models and, in fact, a discrete model, represented by a difference equation, is an intermediate step in deriving a differential equation. Furthermore, since most interesting differential equations cannot be solved explicitly, we have to resort to numerical methods which, in fact, reduce differential equations to difference equations. Interestingly, these often are not the same difference equations which we used earlier in the modelling process to derive the differential equations!

Thus, an important question is whether discrete and continuous models are equivalent in the sense that they describe (approximately) the same dynamics of the modelled process.

5.1 Discretizing differential equations

There are several ways of discretizing differential equations. We shall discuss two commonly used methods.

5.1.1 The Euler method

It is standard in numerical practice to replace the derivative by the difference quotient:

$$\frac{df}{dt} \approx \frac{f(t + \Delta t) - f(t)}{\Delta t}.$$

Then, for instance, the exponential growth equation $N' = rN$, can be approximated by

$$N(t + \Delta t) \approx N(t) + rN(t)\Delta t$$

or, introducing, for a fixed t, the notation $n_k = N(t + k\Delta t)$, by

$$n_{k+1} \approx n_k + rn_k\Delta t.$$

This is a difference equation yielding an approximate value of N at $t+k\Delta t$, $k = 1, 2, \ldots$, provided the initial value at $k = 0$ is given. Note, however, that here we do not have any guarantee that n_k is equal to the true value of N at $t + k\Delta t$ at any time step.

5.1.2 The time-one map

Another method is based on the observation that solutions of autonomous differential equations display the so-called semigroup property: if $x(t, x_0)$ is the flow of the autonomous Cauchy problem

$$x' = g(x), \qquad x(0) = x_0, \tag{5.1}$$

then $x(t_1 + t_2, x_0) = x(t_1, x(t_2, x_0))$. In other words, the process can be stopped at any time and restarted, using the final state of the first time interval as the initial state of the next one, without changing the final output. The semigroup property is sometimes referred to as the *causality property*. Using it we can write

$$x((n + 1)\Delta t, x_0) = x(\Delta t, x(n\Delta t, x_0)). \tag{5.2}$$

This amounts to saying that the solution after $n + 1$ time steps can be obtained as the solution after one time step with the initial condition given by the solution after n time steps. In other words, denoting $x_n = x(n\Delta t, x_0)$, we have

$$x_{n+1} = f_{\Delta t}(x_n),$$

where by $f_{\Delta t}$ we denote the operation of getting the solution of the Cauchy problem (5.1) at $t = \Delta t$ with the initial condition which appears as the argument of $f_{\Delta t}$.

We note that, in contrast to the Euler method, the time-one map method is exact; that is, $x_{n+1} = x(n\Delta t, x_0)$. However, its drawback is that we have to know the solution of (5.1) and thus its practical value is limited.

To simplify the following discussion we shall set $\Delta t = 1$.

5.1.3 Discrete and continuous exponential growth

Let us consider the Cauchy problem

$$N' = rN, \qquad N(0) = N_0,$$

the solution of which is given by $N(t) = N_0 e^{rt}$, and compare it with the discrete model (2.11). As we have seen above, the Euler discretization yields $n_{k+1} - n_k = rn_k$, the solution of which is given by $n_k = (1 + r)^k N_0$. Clearly, the Euler approximation $(n_k)_{k \in \mathbb{N}_0}$ does not match the solution to the continuous problem $N_0 e^{rt}$ at $t = 1, 2, \ldots$. However, their qualitative behaviour is similar, as both grow at an exponential rate and can be mapped to each other by adjusting the growth rates.

On the other hand, consider the time-one discretization which amounts to assuming that we take census of the population in evenly spaced time moments $t_0 = 0, t_1 = 1, \ldots, t_k = k, \ldots$, so that

$$N(k) = e^{rk} N_0 = (e^r)^k N_0.$$

Comparing this equation with (2.12), we see that it corresponds to the discrete model (2.11) with net growth rate $R = e^r$. Thus, if we observe a continuously growing population in discrete time intervals and the observed (discrete) net growth rate is R, then the real (continuous) growth rate is given by $r = \ln R$.

5.1.4 Logistic growth in discrete and continuous time

Consider the logistic differential equation

$$y' = ay(1 - y), \qquad y(0) = y_0. \tag{5.3}$$

The Euler discretization (with $\Delta t = 1$) gives

$$y_{n+1} = (1 + a)y_n \left(1 - \frac{ay_n}{1 + a}\right), \tag{5.4}$$

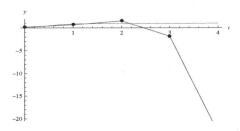

Figure 5.1 Comparison of solutions to (5.3) and (5.4) with $a = 4$.

which is a discrete logistic equation. We have already solved (5.3) and we know that its solutions monotonically converge to the equilibrium $y = 1$. However, if we plot solutions to (5.4) with, say, $a = 4$, we obtain the picture presented in Fig. 5.1. Hence, in general, it seems unlikely that we can use the Euler discretization as an approximation to the continuous logistic model.

Let us, however, write down the complete Euler scheme:

$$y_{n+1} = y_n + a\Delta t y_n (1 - y_n), \qquad (5.5)$$

where $y_n = y(n\Delta t)$ and $y(0) = y_0$. Then

$$y_{n+1} = (1 + a\Delta t)y_n \left(1 - \frac{a\Delta t}{1 + a\Delta t} y_n\right)$$

and the substitution

$$x_n = \frac{a\Delta t}{1 + a\Delta t} y_n \qquad (5.6)$$

reduces (5.5) to

$$x_{n+1} = \mu x_n (1 - x_n), \qquad (5.7)$$

where $\mu = 1 + a\Delta t$. Thus the parameter μ, which controls the long-term behaviour of solutions to the discrete equation (5.7) – see Section 4.2.4, depends on Δt and, by choosing a suitably small Δt, we can get solutions of (5.7) to mimic the behaviour of solutions to (5.3). Indeed, by taking $1 + a\Delta t \leq 3$, we obtain the convergence of solutions x_n to the equilibrium $x^* = a\Delta t/(1 + a\Delta t)$ which, reverting to (5.6), gives as the solution to (5.3) the discrete approximation y_n, which converges to 1. However, as seen in Fig 5.2, this convergence is not monotonic, hence the approximation

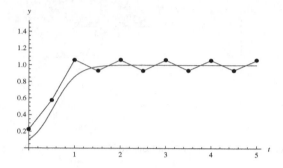

Figure 5.2 Comparison of solutions to (5.3) with $a = 4$ and (5.7) with $\mu = 3$ ($\Delta t = 0.5$).

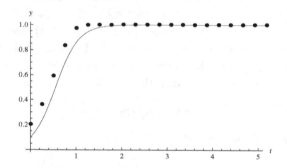

Figure 5.3 Comparison of solutions to (5.3) with $a = 4$ and (5.7) with $\mu = 2$ ($\Delta t = 0.25$).

is rather poor. This can be remedied by taking $1 + a\Delta t \leq 2$, in which case the qualitative features of $y(t)$ and y_n are the same, see Fig. 5.3.

These features of the discrete logistic model can, to a certain extent, be explained by interpreting it as a game between the population and the environment, in which the response of the environment to the population size y_n comes only after the full time step, resulting in the population of y_{n+1}. It is then natural to expect that the system is more likely to lose stability if the response times are long. On the other hand, in the continuous logistic model the responses are instantaneous, resulting in its monotonic and smooth behaviour.

We note that the above problems can also be resolved by introducing the so-called non-standard difference schemes consisting of replacing the derivatives and/or nonlinear terms by more sophisticated expressions which, though equivalent when the time step size goes to 0, nevertheless produce a qualitatively different discrete picture. In the case of the logistic equation such a non-standard scheme can be constructed by replacing y^2 not by y_n^2 but by $y_n y_{n+1}$:

$$y_{n+1} = y_n + a\Delta t(y_n - y_n y_{n+1}).$$

In general, such a substitution yields an implicit scheme, but in our case the resulting recurrence can be solved for y_{n+1}, producing

$$y_{n+1} = \frac{(1 + a\Delta t)y_n}{1 + a\Delta t y_n}$$

in which we recognize the Beverton–Holt equation (2.15) with $R_0 = 1 + a\Delta t$ (and $K = 1$). We have seen in Section 2.4.1 that $(y_n)_{n \in \mathbb{N}_0}$ monotonically converges to an equilibrium and, as we shall see below, it exactly follows the solution of the continuous logistic equation. In the spirit of the game interpretation of the model, discussed above, this stability can be attributed to the fact that the environment response is based on the input $y_n y_{n+1}$ combining the previous and the current time instants, in contrast to y_n^2 in the case of the Euler discretization above.

We complete this section by deriving the time-one map discretization of (5.3). In this situation, the solution (3.20) is

$$y(t) = \frac{y_0 e^{at}}{1 + (e^{at} - 1)y_0}$$

which, upon writing $e^a = R_0$, gives the time-one map

$$y(1, y_0) = \frac{y_0 R_0}{1 + (R_0 - 1)y_0}.$$

Defining $y_n = y(n, y_0)$ (remember $\Delta t = 1$), we obtain

$$y_{n+1} = \frac{y_n R_0}{1 + (R_0 - 1)y_n},$$

in which again we recognize the Beverton–Holt model with the intrinsic growth rate related to the continuous growth rate in the same way as in the exponential growth equation.

5.2 Discrete equations in continuous time models

In the previous section we saw that often it is difficult to describe processes occurring in continuous time using difference equations. This is because instantaneous changes in a continuous process can accumulate in an irregular manner between consecutive observation times and result in large errors. On the other hand, difference equations usually are easier to handle. Here we shall describe two situations in which continual changes yield to a discrete description. First we discuss models with periodic coefficients. They allow for a discrete time modelling, provided one is satisfied with only knowing the state of the system at time instants, corresponding to the period of the coefficients. The second case concerns hybrid models, in which we have a continuous repetitive process interspersed with instantaneous events at evenly spaced time intervals.

5.2.1 Discrete models of seasonally changing populations

So far we have considered models in which the laws of nature are independent of time. In most real processes we have to take into account phenomena which depend on time, such as the seasons of the year. Here we use the same modelling principles as in Section 3.3, but with time-dependent birth and death coefficients $\beta(t)$ and $\mu(t)$. We also consider emigration, which also is supposed to be proportional to the total population, and immigration, which is just a given flux of individuals into the system. Then, instead of (3.13), we have

$$N'(t) = (\beta(t) - \mu(s))N(t) - e(t)N(t) + c(t), \qquad (5.8)$$

where e is a (time-dependent) per capita emigration rate and c is the global immigration rate.

Closed systems. Here we are interested in populations in which the coefficients change periodically with the same period, e.g. with the seasons of the year. As we shall see, contrary to naive expectations, in general this assumption does not yield periodic solutions. We start with a closed population; that is, we do not consider emigration or immigration processes. As in Section 3.3, we define

$r(t) = \beta(t) - \mu(t)$ to be the net growth rate of the population and assume that it is a periodic function with a period T. Under this assumption, we introduce the average growth rate of the population by

$$\bar{r} = \frac{1}{T} \int_0^T r(t)dt. \tag{5.9}$$

Hence, let us consider the initial value problem

$$N'(t) = r(t)N(t), \qquad N(t_0) = N_0, \tag{5.10}$$

the solution of which is given by

$$N(t) = N_0 \exp \int_{t_0}^t r(s)ds. \tag{5.11}$$

Since, by the periodicity of r,

$$\int_{t_0}^{t+T} r(s)ds = \int_{t_0}^t r(s)ds + \int_t^{t+T} r(s)ds = \int_{t_0}^t r(s)ds + \bar{r}T,$$

we have

$$N(t+T) = N(t)e^{\bar{r}T}$$

and hence the solution is not periodic. However, we may provide a better description of the evolution by identifying a periodic component in it. In other words, let us try to find what is 'missing' in the function $R(t) := \int_{t_0}^t r(s)ds$ which stops it being periodic. We observe that

$$R(t+T) = \int_{t_0}^{t+T} r(s)ds = \int_{t_0}^t r(s)ds + \int_t^{t+T} r(s)ds = R(t) + \bar{r}T,$$

so that

$$R(t+T) - \bar{r}(t+T-t_0) = R(t) - \bar{r}(t-t_0),$$

and therefore the function $R(t)$, complemented by $-\bar{r}(t-t_0)$, becomes periodic.

Using this result, we can write

$$N(t) = N_0 \exp \int_{t_0}^{t} r(s)ds = N_0 e^{\bar{r}(t-t_0)} Q(t), \qquad (5.12)$$

where

$$Q(t) = \exp \int_{t_0}^{t} r(s)ds - \bar{r}(t - t_0) \qquad (5.13)$$

is a periodic function satisfying $Q(t_0) = 1$.

In particular, if we observe the population in discrete time intervals of length T, we get

$$n_k := N(t_0 + kT) = N_0 e^{\bar{r}kT} Q(t_0) = N_0 [e^{\bar{r}T}]^k,$$

which is the exponential discrete model with the growth rate given by $e^{\bar{r}T}$.

Open systems. Consider next an open population described by

$$N'(t) = r(t)N(t) + c(t), \qquad (5.14)$$

where $r(t) = \beta(t) - \mu(t) - e(t)$ and $c(t)$ are continuous and periodic functions with period T. Let the constant $\bar{r}$ and the periodic function $Q(t)$ be defined as in (5.9) and (5.13). Using the integrating factor, we find the general solution to (5.14) to be

$$N(t) = N(t_0) \exp \left(\int_{t_0}^{t} r(s)ds \right) \qquad (5.15)$$

$$+ \exp \left(\int_{t_0}^{t} r(s)ds \right) \int_{t_0}^{t} \exp \left(- \int_{t_0}^{u} r(s)ds \right) c(u)du.$$

If there is a periodic solution of period T, say $\bar{N}$, there should be an initial condition N_p satisfying $N_p = \bar{N}(t_0) = \bar{N}(t_0 + T)$. Using

(5.15), we obtain

$$N_p = \overline{N}(t_0)$$

$$= \overline{N}(t_0) \exp \left(\int\limits_{t_0}^{t_0+T} r(s)ds \right)$$

$$+ \exp \left(\int\limits_{t_0}^{t_0+T} r(s)ds \right) \int\limits_{t_0}^{t_0+T} \exp \left(- \int\limits_{t_0}^{u} r(s)ds \right) c(u)du.$$

For simplicity we assume $\bar{r} \neq 0$. By (5.13),

$$\overline{N}(t_0) = \frac{e^{\bar{r}T}}{1 - e^{\bar{r}T}} \int\limits_{t_0}^{t_0+T} \frac{e^{-\bar{r}(u-t_0)}c(u)}{Q(u)} du.$$

Let us define $\widehat{N}(t) = \overline{N}(t+T)$, where $\overline{N}(t_0) = N_p$. Then

$$\widehat{N}'(t) = \overline{N}'(t+T) = r(t+T)\overline{N}(t+T) + c(t+T)$$
$$= r(t)\widehat{N}(t) + c(t)$$

and, since $\widehat{N}(t_0) = \overline{N}(t_0 + T) = \overline{N}(t_0) = N_p$, the uniqueness of solutions of linear differential equations yields

$$\overline{N}(t+T) = \hat{N}(t) = \overline{N}(t)$$

for any $t \in \mathbb{R}$. Hence $\overline{N}$ is periodic.

From Section 1.3.2, the general solution to an inhomogeneous linear equation can be expressed as a sum of an arbitrary solution of the inhomogeneous equation and the general solution of the homogeneous equation. Hence, since $\overline{N}$ is a solution of the inhomogeneous equation (5.14) and the general solution of its homogeneous version is given (5.12), we have

$$N(t) = Ke^{\bar{r}(t-t_0)}Q(t) + \overline{N}(t),$$

where $K = N(t_0) - N_p$. Finally

$$N(t) = (N(t_0) - N_p)e^{\bar{r}(t-t_0)}Q(t) + \bar{N}(t). \qquad (5.16)$$

This formula yields, in particular, that if $\bar{r} < 0$, then

$$\lim_{t \to \infty} (N(t) - \overline{N}(t)) = 0;$$

that is, with negative average growth rate, an arbitrary solution is asymptotically periodic.

Exercise 5.1   Find the difference equation satisfied by the population described by (5.16) if it is observed at times kT, $k = 0, 1, \ldots$.

Exercise 5.2   Find an analogous representation of solutions to (5.14), if $\bar{r} = 0$.

5.2.2 Hybrid models

Absorption of drugs. An important process which leads to an exponential decay model is the absorption of drugs from the bloodstream into the body tissues. The significant quantity to monitor is the concentration of the drug in the bloodstream, which is defined as the amount of the drug per unit volume of blood. Observations show that the rate of absorption of the drug, which is equal to the rate of decrease of the above concentration, is proportional to the said concentration. Arguing as in Section 3.2, we see that this concentration, c, satisfies

$$c' = -\gamma c, \qquad (5.17)$$

with γ being the proportionality constant.

We shall consider a process which is a combination of a continuous process of drug absorption with a discrete process of drug injection.

Assume that a dose D_0 of a drug increases the drug's concentration in the patient's body by c_0 and that it is injected at regular time intervals $t = 0, T, 2T, 3T \ldots$. Between the injections the concentration c of the drug decreases according to the differential equation (5.17). It is convenient here to slightly change the notation and denote by c_n the concentration of the drug just after the nth injection; that is, c_0 is the concentration just after the initial (zeroth) injection, c_1 is the concentration just after the first injection, at the time T, etc. We need to find a formula for c_n and to determine whether the concentration of the drug eventually stabilizes.

First, we observe that the process is discontinuous at the injection times so that we have two different values for the value of c around the time of injection: just before and just after (assuming that the injection is done instantaneously). To avoid ambiguity, we denote by C_n the concentration just before the nth injection and by c_n the concentration just after it, in accordance with the notation introduced above. Thus, using (3.3), we see that between the nth and $n+1$th injection the concentration changes according to the exponential law

$$C_{n+1} = c_n e^{-\gamma T},$$

so that over each time interval between injections the concentration decreases by a constant fraction $a = e^{-\gamma T} < 1$. Thus, we are able to write down the difference equation for concentrations just after the $n+1$th injection as

$$c_{n+1} = ac_n + c_0. \tag{5.18}$$

The solution, using (1.5), is given by

$$c_n = c_0 a^n + c_0 \frac{a^n - 1}{a - 1} = -\frac{c_0}{1 - a} a^{n+1} + \frac{c_0}{1 - a}.$$

Since $a < 1$, we immediately obtain that

$$\bar{c} = \lim_{n \to \infty} c_n = \frac{c_0}{1 - a} = \frac{c_0}{1 - e^{-\gamma T}}.$$

Similarly, the concentration just before the nth injection is

$$C_n = c_{n-1} e^{-\gamma T} = e^{-\gamma T} \left(\frac{c_0}{e^{-\gamma T} - 1} e^{-\gamma T n} + \frac{c_0}{1 - e^{-\gamma T}} \right)$$

$$= \frac{c_0}{1 - e^{\gamma T}} e^{-\gamma T n} + \frac{c_0}{e^{\gamma T} - 1}$$

and, in the long run, $\underline{c} = \lim_{n \to \infty} C_n = \frac{c_0}{e^{\gamma T} - 1}$.

For example, using $c_0 = 14\,\text{mg/l}$, $\gamma = 1/6$ and $T = 6$ hours, we see that after a long series of injections, the maximal concentration, attained immediately after injections, will stabilize at around $22\,\text{mg/l}$. The minimal concentration, just before the injections, will stabilize at around $\underline{c} = 14/e - 1 \approx 8.14\,\text{mg/l}$. This is illustrated in Fig. 5.4.

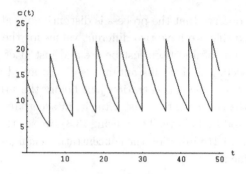

Figure 5.4 Long time behaviour of the concentration $c(t)$.

Population with discrete breeding seasons. Consider a population which reproduces once a year and reproductive season is of a negligible length. Let P_n be the population size in the nth year immediately after the reproductive season. During the year, outside the reproductive season, the population only is subjected to mortality and then the size N of the population obeys the equation

$$N' = -\mu(N)N = -(\mu_0 + \mu_1 N)N, \quad N(0) = P_n,\ \mu_0, \mu_1 > 0,$$

see Exercise 3.2, where t denotes the time that has elapsed since the end of the previous reproductive season (with one year as the time unit). Thus, $N(1)$ gives the population after one year, immediately before the reproductive season. Then, after the reproductive season, the population enters the next year with $P_{n+1} = R_0 N(1)$ individuals, where R_0 is the intrinsic growth rate. To solve the above equation, we separate variables and, integrating, we obtain

$$-t = \int_{P_n}^{N(t)} \frac{ds}{s(\mu_0 + \mu_1 s)} = \frac{1}{\mu_0 \mu_1} \ln \frac{N(t)(\mu_0 + \mu_1 P_n)}{(\mu_0 + \mu_1 N(t))P_n}.$$

Exponentiating and solving for $N(t)$ gives

$$N(t) = \frac{\mu_0 P_n}{\mu_0 e^{\mu_0 \mu_1 t} + \mu_1 P_n (e^{\mu_0 \mu_1 t} - 1)}$$

and, using the reproduction law,

$$P_{n+1} = R_0 N(1) = \frac{R_0 \mu_0 P_n}{\mu_0 e^{\mu_0 \mu_1} + \mu_1 P_n (e^{\mu_0 \mu_1} - 1)}$$

which again is the Beverton–Holt model (2.15).

Exercise 5.3   Find the difference equations for P_n if:

(a) $\mu(N) = \mu_0$;
(b) $\mu(N) = -(\mu_0 + \mu_1 N^\theta)$, $\theta > 0$.

5.3 Stability of differential and difference equations

Let us consider a phenomenon in a static environment which can be described in both continuous and discrete time. In the first case we have the (autonomous) differential equation

$$y' = f(y), \qquad y(0) = y_0, \qquad (5.19)$$

and, in the second case, the difference equation

$$y_{n+1} = g(y_n), \qquad (5.20)$$

with the initial datum y_0. In all considerations of this section we assume that both f and g are sufficiently regular functions so as not to have any problems with the existence of solutions, their uniqueness etc.

First we note that, while in both cases y is the number of individuals in the population, the equations (5.19) and (5.20) refer to two different aspects of the process. In fact, while (5.19) describes the (instantaneous) rate of the change of the population's size, (5.20) gives the size of the population after each cycle. Thus, to make the comparison easier, (5.20) should be written as

$$y_{n+1} - y_n = -y_n + g(y_n) =: \bar{f}(y_n), \qquad (5.21)$$

which would describe the rate of change of the population size per unit cycle. However, typically, difference equations are written and analysed in the form (5.20).

First let us consider differential equations. From Theorem 4.3, it follows that if f has several equilibrium points, then the stationary solutions corresponding to these points divide the (t, y)-plane into

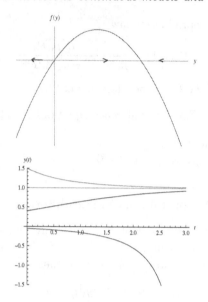

Figure 5.5 Monotonic behaviour of solutions to (5.19) depends on the right hand side f of the equation.

strips such that any solution always remains confined to one of them, see Fig. 5.5.

Remark 5.4 If the reader is familiar with the language of the phase space and orbits, then the last statement can be expressed by saying that the phase space is the real line $\mathbb{R}$, divided by the equilibrium points and thus the orbits are open segments (possibly stretching to infinity) between the equilibria.

Furthermore, we observe that if $f(y) > 0$, then the solution $y(t)$ is increasing at any point t, when $y(t) = y$; conversely, $f(y) < 0$ implies that the solution $y(t)$ is decreasing when $y(t) = y$. This also implies that any equilibrium point y^* with $f'(y^*) < 0$ is asymptotically stable and it is unstable if $f'(y^*) > 0$. In particular, there are no stable equilibria which are not asymptotically stable.

If we now look at the difference equation (5.20), then firstly we note some similarities. Equilibria are defined as $g(y) = y$, (or $\bar{f}(y) = 0$) and, while in the continuous case we compared f with zero, in the discrete case we compare $g(x)$ with x: $g(y) > y$ means

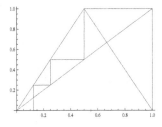

Figure 5.6 Eventual equilibrium $x = 1/8$ for the tent map.

that $y_{n+1} = g(y_n) > y_n$ so that the iterates are increasing, while if $g(x) < x$, then they are decreasing. Also, the stability of equilibria is characterized in a similar way: if $|g'(y^*)| < 1$, then y^* is asymptotically stable and if $|g'(y^*)| > 1$, then y^* is unstable. In fact, if $g'(y^*) > 0$, then we have exact equivalence: y^* is stable provided $\bar{f}'(y^*) < 0$ and unstable if $\bar{f}'(y^*) > 0$. Indeed, in such a case, if we start on one side of the equilibrium y^*, then no iteration can overshoot this equilibrium since for, say, $y < y^*$ we have $g(y) < g(y^*) = y^*$. Thus, as in the continuous case, the solutions are confined to the intervals between successive equilibria.

However, the similarities end here, as the dynamics of difference equations is much richer than that of the corresponding differential equations.

First, unlike in Theorem 4.3(i), a solution of a difference equations can reach an equilibrium in a finite time, as demonstrated in Example 5.5. The points which can reach an equilibrium in finite time are called *eventual equilibria*.

Example 5.5 Consider the tent map T, introduced in Exercise 4.16. As we know, there are two equilibrium points, 0 and 2/3. Looking for eventual equilibria is not as simple. Taking $x_0 = 1/8$, we find $x_1 = 1/4$, $x_2 = 1/2$, $x_3 = 1$ and $x_4 = 0$, and hence 1/8 (as well as $1/4, 1/2$ and 1) are eventual equilibria, see Fig. 5.6. It can be checked that all points of the form $x = n/2^k$, with $n, k \in \mathbb{N}$, satisfying $0 < n/2^k < 1$, are eventual equilibria.

Further, recalling Remark 4.17, we see that if $-1 < g'(y^*) < 0$, then the solution can overshoot the equilibrium and create damped oscillations towards it, whereas in any autonomous scalar

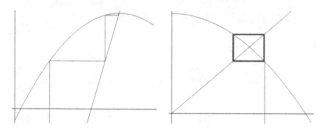

Figure 5.7 Change of the type of convergence to the equilibrium
from monotonic if $0 < g'(y^*) < 1$ to oscillatory for $-1 < g'(y^*) < 0$

differential equations a reversal of the direction of motion is impossible. Also, as we have seen, difference equations may have
periodic solutions which are precluded from occurring in the continuous case. Finally, no chaotic behaviour can occur in scalar
differential equations (partly because they do not admit periodic
solutions, an abundance of which is a sign of chaos). In fact, it can
be proved, see e.g. (Hirsch et al., 2004), that chaos in differential
equations may occur only if the dimension of the state space is at
least 3.

References

M. Braun, *Differential Equations and Their Applications. An Introduction to Applied Mathematics,* 3rd ed., Springer-Verlag, New York, 1983.

N. F. Britton, *Essential Mathematical Biology,* Springer, London, 2003.

R. Courant, F. John, *Introduction to Calculus and Analysis I,* Springer, Berlin, 1999.

S. Elaydi, *An Introduction to Difference Equations,* 3rd ed., Springer, New York, 2005.

W. Feller, *An Introduction to Probability Theory and Its Applications,* 3rd ed., John Wiley & Sons, Inc., New York, 1968.

A. Friedman, W. Littman, *Industrial Mathematics,* SIAM, Philadelphia, 1994.

P. Glendinning, *Stability, Instability and Chaos: an Introduction to the Theory of Nonlinear Differential Equations,* Cambridge University Press, Cambridge, 1994.

M. W. Hirsch, S. Smale, R. L. Devaney, *Differential Equations, Dynamical Systems, and an Introduction to Chaos,* Elsevier (Academic Press), Amsterdam, 2004.

K.T. Holland, A. W. Green, A. Abelev, P.J. Valent, Parametrization of the in-water motions of falling cylinders using high-speed video, *Experiments in Fluids,* **37**, (2004), 690-700.

J. C. Robinson, *Infinite-Dimensional Dynamical Systems,* Cambridge University Press, Cambridge, 2001.

B. J. Schroers, *Ordinary Differential Equations: a Practical Guide,* Cambridge University Press, Cambridge, 2011.

S. H. Strogatz, *Nonlinear Dynamics and Chaos,* Addison-Wesley, Reading, 1994.

H. R. Thieme, *Mathematics in Population Biology,* Princeton University Press, Princeton, 2003.

Index

Printed in the United States
By Bookmasters